SPATIALIZED ISLAMOPHOBIA

This book demonstrates the spatialized and multi-scalar nature of Islamophobia. It provides ground-breaking insights in recognizing the importance of space in the formation of anti-Muslim racism.

Through the exploration of complementary data, both from existing quantitative databases and directly from victims of Islamophobia, applied in two important European capitals – Paris and London – this book brings new materials to research on Islamophobia and argues that Islamophobia is also a spatialized process that occurs at various interrelated spatial scales: globe, nation, urban, neighbourhood and body (and mind). In so doing, this book establishes and advances the new concept of 'Spatialized Islamophobia' by exploring global, national, urban, infra-urban, embodied and emotional Islamophobias, as well as their complex interrelationships. It also offers a critical discussion of the geographies of Islamophobia by pointing out the lack of geographical approaches to Islamophobia Studies. By using self-reflexivity, the author raises important questions that may have hampered the study of 'Spatialized Islamophobia', focusing in particular on the favoured methodologies, which too often remain qualitative, as well as on the whiteness of the discipline of Geography, which can disrupt the legitimacy of a certain knowledge.

The book will be an important reference for those in the fields of Human Geography, Sociology, Politics, Racial Studies, Religious Studies and Muslim Studies.

Kawtar Najib is a social and urban geographer with research interests in social inequality and religious discrimination using both quantitative and qualitative methods. She was a Marie Curie Fellow at Newcastle University where she was the principal researcher of the SAMA project (Spaces of Anti-Muslim Acts), funded by the European Commission, which highlights the impact of Islamophobic discrimination on space and people. She earned her PhD in Geography and Planning at the University of Franche-Comté (France) on socio-spatial inequality and residential segregation in urban neighbourhoods. Her research explores more broadly issues of social and spatial justice.

Routledge Studies in Human Geography

This series provides a forum for innovative, vibrant, and critical debate within Human Geography. Titles will reflect the wealth of research which is taking place in this diverse and ever-expanding field. Contributions will be drawn from the main sub-disciplines and from innovative areas of work which have no particular sub-disciplinary allegiances.

For more information about this series, please visit: www.routledge.com/Routledge-Studies-in-Human-Geography/book-series/SE0514

SPATIALIZED ISLAMOPHOBIA

Kawtar Najib

Routledge
Taylor & Francis Group

LONDON AND NEW YORK

First published 2022
by Routledge
2 Park Square, Milton Park, Abingdon, Oxon OX14 4RN

and by Routledge
605 Third Avenue, New York, NY 10158

Routledge is an imprint of the Taylor & Francis Group, an informa business

© 2022 Kawtar Najib

British Library Cataloguing-in-Publication Data
A catalogue record for this book is available from the British Library

Library of Congress Cataloging-in-Publication Data
A catalog record has been requested for this book

ISBN: 978-0-367-89478-8 (hbk)
ISBN: 978-1-032-11118-6 (pbk)
ISBN: 978-1-003-01942-8 (ebk)

DOI: 10.4324/9781003019428

Typeset in Galliard
by Taylor & Francis Books

To my parents, and especially my mother
(whose dream it has been that I write a book one day)

And to the people who seek real social justice
(even against their own interests and privileges)

Contents

Illustrations

Figures

Tables

Box

Foreword

Islamophobia takes varying forms and presents itself in diverse ways in different places. Geography matters: space plays a complex role in relation to Islamophobia. However, Islamophobia is primarily studied in social science disciplines other than Geography; the tendency in research about Islamophobia is to overlook or side-step the ways in which space might matter.

In this important book, Kawtar Najib foregrounds the role of space when studying Islamophobia and sets out how we can move from Critical Islamophobia Studies towards a critical geography of Islamophobia and anti-Muslim racism. Kawtar builds upon the original data that she collected from victims of Islamophobia in London and Paris to set out what a Spatialized Islamophobia looks like. This study was funded through the award of a prestigious individual fellowship to Kawtar from the Marie Sklodowska-Curie Actions of the European Commission and was hosted at Newcastle University in North East England.

This study is innovative not only in the attention it gives to the lived experiences of those who have experienced Islamophobia, but also in how it combines this with the spatial mapping of Islamophobic incidents to explore how they are concentrated or not in specific neighbourhoods. In doing so, this text makes a crucial point about the strength of adopting a combination of quantitative and qualitative methods in promoting a critical understanding of Spatialized Islamophobia.

Kawtar also reflects critically on her position as a racially and ethnically minoritized Muslim woman working on issues of inequality and Islamophobia, while negotiating the colonial, elitist and exclusionary spaces of academia, and in particular French academia. In doing so, she raises important questions about the whiteness of the discipline of Geography and calls for the inclusion of ethnic and religious minorities within the discipline.

Overall, the significant intervention included in the pages of this book establishes the new concept of Spatialized Islamophobia. Specific attention is paid to global, national, urban, infra-urban, embodied and emotional Islamophobias and the complex ways in which these are interconnected and mutually reinforcing. This is a 'must read' for any scholar or student interested in Islamophobia, racism and inequality in the social sciences, as well as scholars and activists interested in challenging whiteness, discrimination and exclusionary practices in the contemporary academy.

Professor Peter Hopkins
School of Geography, Politics and Sociology
Newcastle University
England, UK
January 2021

Acknowledgements

First of all, I would like to thank my friend Robert Beshara, who strongly encouraged me to write this book and gave me support and help whenever I needed it. A big thank you goes to my friend Josephine Ellis, who read and revised all the chapters; your constant presence means a lot to me. Thanks to the Routledge editors, particularly Faye Leerink and Nonita Saha, and the two reviewers for their crucial comments, especially one.

Many thanks are due to Peter Hopkins for his help and feedback, and to Claire Hancock for our discussions on Geography, Islamophobia, positionality and French academia. I also thank Alison Williamson for the important proofreading.

Finally, I would like to thank all the people (mainly friends) who have supported me throughout this whole year of writing, especially Sarah, Hamza and Nadège.

Acknowledgement of funding

The research conducted for this book was funded by the European Union's Horizon 2020 Marie Sklodowska-Curie programme (European Commission, grant agreement no. 703328) [Horizon 2020-MSCA-IF-2015].

Permissions

Najib, K., & Hopkins, P. (2019). Veiled Muslim women's strategies in response to Islamophobia in Paris. *Political Geography*, 73: 103–111. Copyright Elsevier © 2019. doi:10.1016/j.polgeo.2019.05.005

Najib, K. (2020). Spaces of Islamophobia and Spaces of Inequality in Greater Paris. *Environment and Planning C: Politics and Space*, 1–20. Copyright Sage © 2020. doi:10.1177/2399654420941520

1 Introduction: A spatialized definition of Islamophobia

Introduction

This book has a clear purpose: to show the spatialized nature of Islamophobia. Islamophobia often refers to systemic racism against Muslims and the lived experience of discrimination against people who are perceived as Muslim. It is commonly understood as a process of racialization and Othering that essentializes and homogenizes a group of highly diverse individuals under a single religious attribute (Halliday, 2003; Naber, 2008; Allen, 2010; Sayyid & Vakil, 2010; Scott, 2007). This book contributes to existing definitions by arguing that Islamophobia is also a spatialized process that occurs at various interrelated scales and by providing explanations of how Islamophobia is spatially defined and managed.

Often, in the cases of critical treatments of racism, the geographical component is less explored, yet here I explain how important Geography is to the study of this form of anti-Muslim racism. A geographical lens on Islamophobia shows the need for an increased recognition of 'space' in Islamophobia Studies, primarily by mapping evolving forms of Islamophobia across various spatial contexts while considering critical theories about belonging, identity and intersectionality. I also engage in discussion on the critical geography of Islamophobia in the very last chapter of this book, pointing out the lack of geographical approaches to Islamophobia Studies. I raise important questions about the selected methodological approach, which too often remains qualitative (Hopkins, 2009), as well about the whiteness of the discipline of Geography and the role of racism within it (Kobayashi, 1994).

This book develops the advanced concept of 'Spatialized Islamophobia' by arguing that Islamophobia occurs at various scales (globe, nation, urban, neighbourhood, body (and mind)) that are connected (Najib & Teeple Hopkins, 2020). It provides a specific spatial definition and details important distinctions considering the different scales at which this phenomenon operates (from the global to the micro-local). This glocal approach shows how the global fear of Islam can have a direct impact on, for example, the way that Muslim bodies feel constrained in their everyday life (sense of security, mobility, behaviours, etc.) (Najib & Hopkins, 2019; Pain & Staeheli, 2014).

Indeed, according to the scale under study, Islamophobia changes its contours, its effects, its intensity and its functioning, so that it is difficult to believe that it is

DOI: 10.4324/9781003019428-1

the same phenomenon or even that the scales are distinct from each other. This introduction does not strictly present each scale separately, because the scales interrelate, intersect and reflect upon each other in such a complex way that it is vital not to isolate each for explicit consideration from the general understanding of Spatialized Islamophobia. I therefore describe in this introduction the inter-relationships between the spatial scales in order to understand how multi-scalar Spatialized Islamophobia is. Since the following chapters set out the salient features of the three main levels of scales (global and national in Chapter 2; urban and infra-urban in Chapter 3; embodied and emotional in Chapter 4), this first chapter emphasizes how Islamophobia is connected between scales, from world-wide Islamophobia (through international representations and policies) to intimate Islamophobia (within the family, for example).

But before studying the main purpose of this book, I discuss here, for the first time in my career, my own positionality in order to reflect critically on the study of Spatialized Islamophobia. Building on theories of feminist epistemology (Haraway, 1988; Rose, 1993; hooks, 2000; Harding, 1986, 2004a; Dorlin, 2008), this first section revealing my positionality follows the idea that any knowledge is situated, and those who claim that their knowledge comes from a position of total neutrality are ignorant of this position (Husson, 2014; Chiseri-Strater, 1996; Bourdieu, 2001; Harvey, 2001). Relying especially on feminist standpoint theory, first articulated by the philosopher Sandra Harding (1986), the next pages are dedicated to situating myself and detailing my academic background that led me to write this book. Being directly affected by discrimination reinforces, for me, the importance of discussing it. Harding (1986, 2004b) argues that the best way to understand oppression is to study it from the perspective of the oppressed. Therefore, as a Muslim woman of colour, as a French geographer of North African descent, I am well situated to study the spatial dimension of Islamophobia and engage in a possible discussion on a critical geography of Islamophobia.

A difficult start: studying Islamophobia in a hostile and subversive context

This book is not only the logical continuation of the SAMA (Spaces of Anti-Muslim Acts in Paris and London)[1] research project, funded by the European Commission, and the special issue on Geographies of Islamophobia[2] in *Social and Cultural Geography*. It also results from my own academic experience, when it became evident to me that I was undoubtedly experiencing differential treatment because of who I am. Indeed, the journey has not been easy and has most of the time led me to make important decisions that may have a direct and risky impact on my own professional career development. These necessary sacrifices prompt an important critical reflection on the state of academia and the functioning of scientific research on a global scale, and especially in France.

The biographical details of how this book came to be written are relevant to its reading, as they serve to explain elements of its construction that might otherwise appear too innovative or difficult to grasp. The striking feature was to stop writing articles in French and applying for various fellowships in France, and to become more involved in the Anglo-American academic world where research on race, ethnicity, religion and discrimination is more advanced (Staszak, 2001) and accessible to racially minoritized researchers. It is becoming increasingly clear that in France there are significant reactionary forces on the march to eliminate certain themes, in particular the study of Islamophobia, and systematically to question the skills, contribution and quality of research by racialized[3] and Muslim researchers. It is at present difficult to prove this statement, but my reflections began within French academia (Mondoloni, 2019; Leprince, 2019; Hajjat, 2020) with many testimonials from researchers (from ethnic-minority groups or not), as well as my own professional experience in the French academic world.

That said, one specific and important example clearly illustrates this. A conference on 'Fighting against Islamophobia, an issue of equality' was cancelled in 2017 at the University of Lyon II (a university deemed to be rather non-reactionary) under pressure from far-right and *laïcardes*[4] French organizations, denouncing the participation of suspicious activist contributors (Burlet, 2017). It is indeed difficult to study the theme of Islamophobia in France; which seriously threatens academic freedom. This will surely get worse with the recent Senate amendment (put in place just after the horrific beheading of a teacher in France in October 2020) that redefines the limits of academic freedom in accordance with Republican values (Lentin et al., 2020).

This threat of academic authoritarianism mainly targets researchers who work on racial and decolonial studies, and can be read in different ways. French researchers may be subject to serious criticisms that go so far as to call into question their academic position. French researchers may have to learn English, write in English and disseminate their work internationally rather than nationally, or even move to an Anglo-American country. French researchers may need to seek foreign funding rather than to solicit French institutions, and so on. These constraints, in addition, are strongly associated with several 'prohibitions' imposed by the doxa of French academia's rational objectivity.

First, in France, being objective and neutral is what the traditional scientific culture requires of researchers in order that their works are considered to be of a guaranteed quality. French academia, and especially the discipline of Geography, explains that serious knowledge can only be produced by researchers who are outsiders to what they describe (Weber, 1959; Grasland, 2012; Morelle & Ripoll, 2009). This obviously excludes researchers who are directly concerned, those who are activists and those who have clear political activities. But this implies, above all, that researchers must not state their personal starting point for engaging with their research area. This calls into question the issue of positionality, which is very little discussed in French Geography. Here, the difference from the Anglo-American academic world is more than significant (Staszak, 2001; Gintrac, 2015).

Second, French academia teaches us not to use the first person singular 'I', even in single-author works, for the sake of attenuating the researcher's personal presence in his or her written productions in order to seek an axiological neutrality (Weber, 1959; Bourdieu, 2000). The idea is not to make value judgements about what the researcher is studying; his or her gaze is supposed to be neutral and objective.

Third, French academia wants researchers to distance themselves from activism and political activities (Hancock, 2016). That is why, in such a context, studies on systemic racism and particularly Islamophobia have been constantly marginalized in social sciences in France. There are no Racial Studies or Religious Studies departments in France, due to the refusal of its constitutional Council to construct any reference system on a racial or religious basis, letting such works notably on France's Islamophobia problem to be conducted by foreign researchers, such as for example the US anthropologist Mayanthi L. Fernando with her book, *The Republic Unsettled: Muslim French and the Contradictions of Secularism* (2014), or the US historian Joan Wallach Scott with her book, *The Politics of the Veil* (2007). French academics in general, and geographers in particular, shy away from any overlap between research and activism (Hancock, 2016). Those who are activists take care to keep their activism as separate as possible from their research in order to be seen as objective researchers. Grasland (2012), in this sense, acknowledges this and agrees that being engaged can damage the scientific objectivity of researchers and transform their axiological neutrality due to subjective motivations.

French academia, and notably French Geography, wants researchers to be outsiders and uninvolved with what they study, as if that guarantees *de facto* a neutral, rational and partial scientific output. Yet being an outsider and not engaged does not necessarily make a researcher less biased than an insider, especially when the research is about discrimination and oppression. In any case, why should we believe that subjective motivations are the sole prerogative of insiders? This injunction to exteriority, instead, raises the question of the dominant knowledge (male, White, heterosexual, etc.) that frames the possibilities of debate. If the researcher should not be a member of any discriminated category, it implies that no serious knowledge about sexism, for example, should come from women and no serious knowledge about racism from racially minoritized researchers. Sometimes, to render the scientific argument more allegedly serious, it may be enough to add the name of a male or a White researcher (Mott & Cockayne, 2017).

Precisely, this reasoning cannot be taken seriously, as it implies that only White heterosexual men, Christian or atheist, and from the middle or privileged classes should work on topics of discrimination, oppression and injustice. Their scientific knowledge would be the only type applicable since they are not discriminated against; however, while they might not be its victims, they are still involved in discrimination since they benefit (directly or indirectly) from it, and therefore have an important interest in maintaining it. They think of themselves as 'critical' (Ahmed, 2012: 179), but their works are more likely to follow a dangerous dominant trend under cover of great values, such as Republican Equality or Secularism, which serves to reaffirm their superiority over their dominated colleagues and oppressed people in general. Because of their position of privilege, they are more likely, themselves, to be the 'oppressor' of oppressed categories. Here, it is dangerous to permit, in the name of objectivity and neutrality, the entire scientific space to be occupied by researchers who may themselves be sexist, racist or Islamophobic.

It is therefore time to take on the scientific space that has hitherto not been given to researchers who want to challenge this reality and those who are directly affected by discrimination. Indeed, it is important to understand what researchers are studying and their starting point for engaging with their research area. Especially in France, there is a need to understand that it is not less objective to make it clear from whence researchers speak; on the contrary, it is an even greater guarantor of objectivity (Haraway, 1988; Harding, 1986, 2004b) because they have been transparent about their positionality, their position as investigators and the situation from whence they speak.

But this implies reflexivity. The reflexive approach is not commonly adopted in France, particularly not within the discipline of Geography (Gintrac, 2015). It is a little more used in French Sociology (Naudier & Simonet, 2011; Nora, 1987; Fossé-Poliak & Mauger, 1985; Bourdieu, 2004), yet needs to be better promoted in order to leave more room for a set of researchers who may feel deeply marginalized within French academia or dare not study the forms of oppression, such as sexism, racism, classism, ageism, sizism, ableism, and so on, that directly affect them. While others can deny the existence of all these '-isms', the oppressed are, by contrast, well positioned to see it and analyse them. The oppressed are able to observe not only the non-individual and random nature of oppression but, above all, its collective and systemic nature (Ahmed, 2012). Feminist researchers have been particularly concerned with the role of reflexivity in the social sciences (Haraway, 1988; Wilkinson, 1988; Stanley & Wise, 1983; Harding, 1986, 2004a), and critical reflection on the part of the researcher has served as a bridge between empirical research and political engagement, notably on social issues (Parker, 1994). Therefore, taking the approach of reflexivity is not easy for those from French academia and studying Geography.

It is during the writing of this book that I personally felt the need to clearly express my positionality, which determines where I stand in relation to others (Merriam et al., 2001). The aim is to go forward into the analysis of my feelings and hostilities and thus clarify my motivation through scientific research, enhancing the integrity of my research and making it more valuable. The standpoint of oppressed people is indeed valuable, since it has undoubtedly something to teach people who do not live with a specific oppression. In sharing my experience, I can therefore, at a personal level, analyse and discuss a phenomenon such as Islamophobia that I have seen with my own eyes. The idea is not to talk about me – which does not necessarily make me comfortable because of my French academic background and training – but rather to expose my experience as part of a discriminatory set of institutional structures that are primarily systemic. Therefore, from my own individual identity, I can speak for a collective identity (Collins, 2002). Harding (1998) explains that those who are directly affected by oppression reach a 'strong objectivity' because of their own experience. With other feminist scientists (Haraway, 1988; Husson, 2014; Chiseri-Strater, 1996), she criticizes the illusion that knowledge can be non-situated and neutral, explaining that there is no neutral knowledge. Insiders in research areas in France may see their skills and findings being minimized and even denied when they should be taken into consideration and valued, since their own experience almost speaks for itself.

Finally, I ask for my scientific standpoint and research findings to be heard. Is this too much to ask? I do not think so. Every standpoint should be allowed to be expressed and discussed (even by the majority and therefore 'dominant' category). I personally think that it is complementary and refreshing to discuss a topic with different people and different standpoints – when it is a constructive discussion and not a controversial confrontation. This ensures respectful conditions for dialogue and debate because, at the end of the day, the real challenge is to work together to challenge racial discrimination, rather than against each other.

Consequently, I have to explain here who I am and contemplate how my background, identity, values, beliefs, attitudes and implications may have played a part in my research in general and, in particular, in this book on Spatialized Islamophobia. First of all, I have always been interested in research topics that focus on the dialogue between societies and urban spaces. Thus, I see myself first and foremost as a social and urban geographer. I have analysed this link in different ways throughout my studies and academic positions. My first research work (in my Masters at the University of Franche-Comté in France) explores the spatial practices of immigrant religious groups (i.e. Muslim and Buddhist populations) and their integration in France. I was already able to see how difficult it was to address this issue of religion and immigration in a country that does not recognize religious and cultural minorities and thus cannot even, for example, sign a charter for their protection.

Therefore, for my doctorate at the same university, I did not want to continue to work on such a topic. I decided on another issue, more related to urban segregation, as this seemed easier to study in France. But it was also difficult, especially within a school of quantitative geography, to study the issue of socio-spatial inequalities and to raise debate about the transparency of ethnic data in France, the existence of ghettos, the extent of racism and the ineffectiveness of government actions, particularly on social mixing and diversity. Indeed, quantitative geography is usually perceived to be too close to the State and at the service of dominant economic interests (Harvey, 1972; Luxembourg, 2016). After seeing how challenging it was for a student like me to conduct such research, I realized that, for me, these were formative experiences. I realized that French Geography – which has little room for critical geography as studied in Anglo-American research (Gintrac, 2012, 2015, 2017) mobilizing the question of race, religion, gender, sexuality and other intersecting identities – was not yet ready for this type of research.

At the same time, I witnessed increasing media coverage and political debate about the visibility of Muslim populations and their difficulties in integrating into European society. I wanted to analyse this issue with a geographical perspective. I started to write a postdoctoral research project on the spaces of anti-Muslim acts, in line with my previous research experiences on religious minority groups and spatial inequality. I applied for a national fellowship with the help of Professor Claire Hancock (from the University of Paris-Est Créteil), one of the very few geographers working on this issue in France, but my project was not accepted. Then I decided to write a version of this project for a European fellowship (the Marie Curie grant, one of the most competitive and prestigious in Europe) with Professor Peter Hopkins of Newcastle University, and it was accepted.

Finally, I knew that my research interests, especially the study of Islamophobia, since they concern themes revolving around the issue of social and spatial justice, religious and territorial discrimination, and social and urban exclusion, would not be fully acknowledged by French scientists and would generate significant criticism. I faced negative comments and attacks by French colleagues, questioning my academic skill and scientific rigour. Some have even publicly denigrated my work on Islamophobia, using specific words like fuzzy, disappointing, not serious... They claim that my work is more militant propaganda that hinders criticism of Islam and does not respect the values of *laïcité*. [5] They clearly wonder how I managed to find a place in an institution dedicated to the production and transmission of knowledge since, according to them, I refer to any condemnation of Islamism as a form of discrimination.

These comments are very different from what my co-authors and I offer in our work; namely, a geographical comparison of anti-Muslim acts in two important European capitals. It is therefore obvious that such a view (unfortunately dominant in France) reflects the great difficulty faced by researchers like me in integrating into French academia. I knew that I must not be distracted by unfounded comments and attacks that aim to discredit my scientific work, which is rewarded abroad. Indeed, in France it is difficult to study such a topic, and in particular to combine Geography and Islamophobia. From this observation, I saw that I had to move to a country such as the United Kingdom to work on this area, as the problem is directly related to my identity that amplifies the level of criticism and tensions that I am exposed to.

Even though I like being referred above all as a geographer due to my academic background, I have nevertheless to state that I am a French woman of colour of North African descent (or Maghrebi, more exactly from Morocco) from a working-class and low-educated family (my father is illiterate, for example), who grew up in a deprived area in a medium-sized city in the east of France. I did not enjoy many privileges, but I had free access to a high level of education thanks to the establishment of free schooling for all in France in the 19th century, and I grew up in a loving and encouraging family, which is a huge privilege barely considered and studied in research on intersectionality and privilege in contemporary societies. Here, it is important to explain that though I am in various stigmatized categories and therefore may be qualified to study oppression, I acknowledge that I am not necessarily able to discuss forms other than those I know, not for example, disability, sexual orientation or size.

More importantly, regarding Islamophobia Studies, I consider myself as a Muslim and religious woman. Indeed, there is an important personal engagement with the material studied since I see myself as a spiritual person who used to work on Islam and Buddhism and who looks to promote social justice in our societies, hence my work on socio-spatial segregation and Islamophobia. Besides, I am familiar with the Muslim community, and especially the French Muslim community, since I can understand Arabic (North African Arabic dialect, to be exact) and can easily refer to and engage in Muslim practices and habits.

When interacting with victims of Islamophobia in the interviews in Paris and London, I revealed my religious identity to those participants who asked me and whenever I considered it to be a useful way to build a trusting relationship. For example, I was able to use Arabic greetings (such as *Assalam alikom*) and Muslim formulas (such as *Insha'Allah* or *Alhamdulillah*) when interacting with interviewees, who were all Muslims, and the vast majority religious. As for interaction with British participants, I felt the need to be (usually) accompanied by a female assistant from a South Asian background (in collaboration with the SAMA project) in order to reassure interviewees and to be able to communicate with them if their English was limited. I was able to understand them and communicate with them, but English is not my primary language.

Positionality is not only how I see myself but also how others see me, through my research. I see myself as a Muslim woman yet, since I do not necessarily look Muslim to everybody, some will not immediately perceive my religious identity while others will assume that I am not religious (or, not religious enough). My research with victims of Islamophobia, who all considered themselves as Muslims, provides important considerations in terms of differences and similarities. Indeed, it is well known, notably in the social sciences, that who you are opens or closes doors and determines who will feel confident to cooperate with you, as well as whether you will feel comfortable with the interviewees or not, and therefore determines the type of discourses that you will be able to collect (Sue, 1993; De Galembert & Belbah, 2004). For example, Sue (1993) has shown that the atmosphere of suspicion and mistrust that many ethnic-minority groups feel toward a White researcher of ethnic matters may be understandable from the perspective of the wider social forces that reflect our society. Thus, I was most of the time considered as an insider, thanks to my skin colour, my name, my partnership with associations that they trust, my knowledge of their religious practices and habits, and my inclusion in religious minority groups in Europe. Sometimes the perceived difference between the research participants (who might, for example, wear a *niqab* (the full-face veil)) and myself was more to do with the ways of understanding and practising the Islamic faith, rather than them considering me as a real outsider.

As for my gender, the victims of Islamophobia in France and in the United Kingdom are in the great majority women (Najib & Hopkins, 2020). I was therefore able to share experiences with them in terms of gender identity. When men participated in the project, the distance was noticeable yet did not represent a significant obstacle. I also managed to meet every single participant in a public space. All these aspects allowed me to create an allied environment rather than a great distance between the interviewees (the researched) and myself (the researcher). But once the interviews started, I have to say that all the barriers that could have been between us progressively disappeared, since the participants basically shared with me an intense and stimulating account of the consequences of

Islamophobia on their everyday life. Some interviews even continued after stopping the recording, in a less formal conversation.

Our positionality as researchers will never be fully understood: how it has manifested during a project, how it is perceived and interpreted by others, and how it has influenced the participants (Rose, 1993; Hopkins, 2009; Gale & Hopkins, 2009). This difficulty can lead us to ignore its influence, as promoted in French research, yet in fact there is no way to do so. It is indeed useful for researchers to think critically about their positionality, as well as the situation of the researched, and to try not to let them think that they are exploited as 'laboratory rats' without being given anything back. As researchers, we need to consider the researched and let them express and share their feelings and experiences as real actors who will help us build our scientific argument. We should not transform them into mere objects of our study. That is why thinking about the methods appropriate to our research is an important process that we need to take seriously, as well as preparation of the interviews in compliance with the ethics requirements. These concerns needed to be addressed and understood in order to conduct ethically sound research.

This work was thoroughly and carefully planned, and it had to secure various approvals before any fieldwork could start. The main ethics guidelines that the SAMA project had to follow were those required by the European Commission, as well as the Ethics Committee of the Faculty of Humanities and Social Sciences of Newcastle University under the surveillance of a project advisor (from the European Commission), a research supervisor and an ethics advisor (from Newcastle University). A researcher needs to be involved and concerned about those researched. If the researcher keeps at too far a distance from the researched and the studied topic, there is an increased risk of a form of voyeurism, and that is why I personally felt the need to show my positionality when needed, as well as my motivation. I showed them my intent to denounce this form of anti-Muslim racism and engage against any form of injustice based on identity characteristics.

Although French academia believes that research needs to disengage from activism, I think that engagement can be part of the scientific process (Bernardie-Tahir & Schmoll, 2012); that is why, for the SAMA project, I worked in collaboration with important non-governmental organizations (NGOs), such as the CCIF (the Collectif Contre l'Islamophobie en France, or Collective against Islamophobia in France) and MEND (Muslim Engagement and Development), which helped me to contact the victims on their database. It required me to follow a set of guidelines and procedures related to the anonymity of the victims, respect for their feelings and a reduction of the sense of intrusion. The interviewees needed to feel confident and at ease, and they all freely consented to participate in the SAMA project.

Finally, upon reading the literature on Islamophobia and discussing the problem with researchers, activists and MPs in international conferences and various meetings, the feeling of wanting to be involved and to denounce this injustice was already important. But it became even stronger after interviewing the victims in Paris and London, some of whom started crying while explaining their experiences. Some had experienced seriously violent situations, such as the young French woman whose aggressor tried to stab her multiple times in front of her seven-year-old child. Clearly, I am not speaking on behalf of the victims of Islamophobia. I just want their voices to be heard and recognized. Indeed, any researcher who works with people who suffer from major difficulties and have undergone important injustice should want to make it their job to recognize the situation and to do something about it (Bernardie-Tahir & Schmoll, 2012; Morelle & Ripoll, 2009), even outside the research world.

If I was told a few years ago that I would expose my positionality, I would not have believed it. Because of my French and secular academic education, it was something that I never considered. Anglo-American academia gives a place to minorities that astonishes French geographers (Staszak, 2001), who work in a context promoting social mixing based on the French Republican Equality model, in contrast to Anglo-American geographers who recognize various heterogeneous groups within their societies and are happy to see these communities living alongside each other rather than coalescing (Collignon, 2010). Since I had direct access to Anglo-American research, reading all these researchers detailing their personal background let me really understand the meaning and value of exposing the researcher's positionality. Indeed, once I clearly understood and exposed my links and motivations regarding this issue, I became better able to balance the emotions from the analysis and build a more solid argument and interpretation. This is a point I want to highlight clearly, even at the risk of seeing my career in French academia totally compromised; instead of protecting my own career, I prefer to stand for justice and thus protect my own integrity.

French geographers may suspect that the substance of my research is biased, but what kind of bias are they thinking about? I do not see it, and if anybody asks me if I am against Islamophobia, then yes, of course I am against Islamophobia, as I am against any form of racism and injustice. Are they concerned because I am a Muslim? Until now, I have never clearly acknowledged this in any written or oral presentation, because France denies the cultural hybridity of immigrants' descendants and teaches us to keep religion to the private sphere. Are French geographers concerned by the potential political and activist affiliations? But I am not a member of any political party or activist association. I may express clear political positions in my works relating to the fight against inequality and discrimination, but I have no direct political and activist functions. Are they concerned because I have an Arabic name and I am a woman of colour? I do not know, but it is deeply disturbing for a researcher like me, who has as much credibility and legitimacy as many researchers, even to think of it.

In the end, it is important to understand that this reflexive work is not narcissistic (Bernardie-Tahir & Schmoll, 2012); on the contrary, it allows me to situate my research, to contextualize my position and engagement, to detail the conditions of the data collection and the methods used and to locate my position and motivations within this. This is a rigorous research process that needs to be better acknowledged in France, especially in Geography, because it is important to understand that stating where I am from and exposing the relationship that I have with this topic is a necessary validation of my scientific approach (as important as clarifying the selected methods, the recruiting process of interviewees, etc.). I know that the reflexive approach is not often used in French Geography, but its potential needs to be exploited in order to allow a new wave of critical researchers to flourish in France, better to serve society.

Finally, this first-time exercise has not been easy for me, since it is also a personal journey that I am sharing with readers. Such transparency is needed to apprehend my interactions with others: researchers, interviewees and the general public. My biographic trajectory is important to understand how I ended up writing this book on Spatialized Islamophobia, but if you read again the title of this book, you will see that my geographical expertise is what this book is really about. Therefore, I continue to consider myself, above all, as a social and urban geographer (no matter what others – researchers or not – may think) who criticizes social and spatial inequalities while trying to make a positive difference. This is why, after laying my cards on the table, after having thought about and explained my positionality and motivations, and contextualized my geographical vision in line with feminist theories in these first pages, I am now fully able as a geographer to give my scientific expertise on the spatialized dimension of Islamophobia in the next pages of this book. So, welcome aboard!

Islamophobia Studies

Islamophobia is not only a human rights issue; it is also the focus of a relatively new field of research, Islamophobia Studies. This field is primarily addressed by social sciences that formulate concepts and categories that are dependent on existing social and power relationships. This section briefly outlines the genealogy of the term 'Islamophobia', which initially began to be theorized as a field of inquiry to a greater extent in social science disciplines other than Geography, notably Sociology, Anthropology and Political Science (Halliday, 2003; Naber, 2008; Garner & Selod, 2015; Allen, 2010; Sayyid & Vakil, 2010; Scott, 2007; Massoumi et al., 2017). The definitions generally emphasize that Islamophobia is rooted in racism and thus describe a form of anti-Muslim racism. It is important to state here that the very concept of Islamophobia has been contested for decades (Lean, 2019), and its terminology itself is sometimes still questioned (Allen, 2010). This book will not question the validation of the term because, in my opinion, the literature has already discussed it in depth. Moreover, a single semantic focus does not always allow us to analyse other aspects of Islamophobia and, most importantly, to think of relevant and specific solutions to combat this anti-Muslim hate concretely.

I acknowledge that Islamophobia is not a perfect concept (just like those such as anti-Semitism, which is less disputed) and has various limitations. This leads me to use also the notion 'anti-Muslim racism'. The main problem with the term of Islamophobia is that it can conflate Islam and Muslims into a single entity and focus only on anti-Islamic rhetoric (Abbas, 2009). Indeed, although previous research has found a correlation between hostility against Islam and the rejection of Muslim populations (Bleich, 2011; Klug, 2012), Islamophobia is not just about

Islam. It is more than that: it also targets the visible markers associated with the Muslim identity.

An important semantic work has already begun, led by many social scientists who have conceptualized it as a form of systemic anti-Muslim racism that targets expressions of Muslimness (Allen, 2010; Sayyid & Vakil, 2010; Carr, 2016). They have tried to describe this phenomenon by referring to a historical contextualization of Islamophobia and intellectual engagement in diverse fields of theoretical analyses, such as racialization, Orientalism, colonialism, imperialism and so forth (Itaoui & Elsheikh, 2018). For example, the work of Edward Said on Orientalism (1978) – which describes how Western discourse represents the figure of the 'Arab' as an exotic and barbarous Oriental – is frequently cited in Islamophobia Studies, and represents one of the primary critical studies on the stigmatization and demonization of Muslim identities. In this sense, the very origin of the word 'Islamophobia' is also related to racist contexts, even if it is contested, provoking competing claims among authors (Allen, 2010). The oldest known references go back to 1910, when the term appeared in the context of French colonialization in Algeria (ibid.) or to 1925 in the French book, *L'Orient vu de l'Occident (The East Seen from the West)* by Étienne Dinet and Sliman B. I. Baamer (as cited in Gresh, 2004).

Whatever the date of its coinage, the use of the term 'Islamophobia' has become more widely associated with the Runnymede Trust reports (1997, 2017), which provide an influential and comprehensive portrait of the many forms of anti-Muslim sentiment and their negative repercussions on the lives of Muslims. The first research works have focused on the demonization of Muslims in the West (Poole, 2002; Parekh, 2000). Progressively, the term has become widely used in academia, and the definition most used by scholars states that Islamophobia appears as a racialization process of individuals that essentializes and homogenizes Muslims, as well as those who are perceived as such, and constructs thinking about Muslims and Islam as Other (Allen, 2010; Sayyid & Vakil, 2010; Garner & Selod, 2015).

Hafez (2018) offers an interesting overview of theoretical debates in the field of Islamophobia Studies and identifies three main schools of thought using this term:

1 *Prejudice studies:* Islamophobia is mainly about Muslims. Here, Islamophobia is simply another reincarnation of the unfortunate trend of bigotry that Jews, Afro-Americans and other populations have had to face throughout history. The dimensions of power and domination are significant and contribute to an understanding of how Islamophobia works through a process of homogenization that discriminates against Muslim populations. Muslims and mainly Western Muslims are suspected and accused of being the 'enemy within' (Fekete, 2009).

2 *Racism and postcolonial studies:* Islamophobia is about the dominant culture, mainly in Western societies. Here, Islamophobia is primarily apprehended in an asymmetric power relationship and makes theoretical links to critical race and postcolonial studies (Said, 1978; Spivak, 1988; Maldonado-Torres, 2007). The very aim of these studies is to criticize power structures that seek to govern the subjects that they have constructed. They do not focus on

individual racism, but rather on the awareness of the racist practices inscribed in our socialization and knowledge.

3 *Studies on decoloniality:* Islamophobia is about bringing the 'Muslim subject' back to a subaltern voice that has no agency and needs to be controlled. Here, the fight against Islamophobia becomes part of a larger global struggle against racialized inequalities and exploitation, and the epistemological struggle that comes with it introduces a post-positivist, post-orientalist and decolonial perspective to overcome the violent hierarchy and domination of the West over the non-West (Grosfoguel, 2011; Sayyid, 2014).

The field of Islamophobia Studies has to deal with these various ideological understandings, but Massoumi et al. (2017), for example, have identified concrete social, political and cultural actors that reproduce ideas that exclude Muslims (such as the State, the far-right, the neoconservative movement, the transnational Zionist movement and assorted liberal groupings, including the pro-war left and the new atheist movement). They refer to these as the 'five pillars of Islamophobia'. Despite these various discussions, Islamophobia is still a useful umbrella term. It is widely used in different studies, whether they focus on the theorization and conceptualization of the phenomenon, its politics and legal system, its representation in the media, its industry, its materialization in terms of concrete discrimination, its Othering and racialization process or its gendered and racialized dimension (Itaoui & Elsheikh, 2018).

The idea of Islamophobia has significant purchase among intellectuals, scholars, influential think-tanks and government departments. Indeed, over the last three decades, Islamophobia has become part of established academic, political, practitioner and societal parlance (Garner & Selod, 2015), and the term is widely understood as a form of anti-Muslim racism. Opposition to the use of this term rarely comes from constructive discussions, instead arising as controversy emanating from the various power structures which may even question its existence (Lean, 2019). Therefore, I leave these sensational polemics and semantic work to others, since this book aims to go beyond these debates by bringing a spatial reading to this anti-Muslim racism, and thus explores the effects that it has on spaces and people.

Islamophobia and its spatial dimension

Understanding the significance of space

While the field of Islamophobia Studies has indeed increased in recent decades, few focus on its geographical dimension. Islamophobia remains, from a general point of view, primarily addressed by sociologists, anthropologists and political scientists, and important international publications and conferences bear witness to this. Yet space, and especially public space, is at the centre of concerns raised by the phenomenon of anti-Muslim rejection. Therefore, 'space' is a privileged actor in the study of Islamophobia and needs to be better recognized, not only in

Islamophobia Studies in general but in particular within the study of geographies of Islamophobia. Indeed, while working on geographies of Islamophobia, I always felt myself to be in an extremely minority position in Geography where the study of Islamophobia is partly marginalized, and within Islamophobia Studies where geographers are rare. That said, in light of increased discrimination against Muslims in public space, discussions around the spatial dimension of Islamophobia have begun to develop (Najib & Teeple Hopkins, 2020). Islamophobia manifests itself differently according to space, and geographers are typically well positioned to study the spatial aspect of Islamophobia and in particular its extent, divisions/connections, organization, nature and content.

First of all, 'space' is multidimensional and has multiple meanings (Harvey, 1973). Of course, something as complicated as the 'space system' requires an extensive analytical framework, but here I will provide only a general understanding of space and its related notions (such as scale and place) that can be found in introductory Geography books. Space is a polysemic term that may be used in other disciplines (such as in Astronomy and Mathematics) but, in Geography, space represents a delimited portion of the terrestrial extent, therefore one that is observable and mappable. The neutral and broad definition of space (Gieseking & Mangold, 2014) usually pushes geographers to add an adjective to qualify it; this is why we can talk about urban spaces, residential spaces, wooded spaces, cultural spaces, everyday spaces, perceived spaces, and so on.

Space does not depend only on its natural environment and material or immaterial heritage; it is also developed, managed, modelled and built by societies and their activities. Indeed, the individual is 'situated' in a specific environment that conditions the appreciation of his or her own development and well-being. Society and space are correlated: one is continually producing and changing the other (Little, 2014; Valentine, 2001). Therefore, it is important to recognize that space (and also time) is fundamental to understanding society, and *vice versa*. Social geographers focus on the implications of this relationship for social identities, social reproduction, social inequalities and social justice; and that is why, as a social geographer, I am interested in how geographies of Islamophobia work both in large scales and in particular places.

Scale and place are also terms related to space that are often used in Geography (Massey, 1991). Scale primarily refers to the ratio of a measurement on a map and the corresponding distance on the ground. It describes the size, scope or extent of something compared with something else, and is a measure using a geographical reference system. Place defines more precisely the location and the phenomenon that one describes. Massey (1991) gave an important meaning to place by explaining that processes operating at a range of spatial scales come together in diverse ways to create experiences of place. These processes can create material inequalities between and within places. In the study of Spatialized Islamophobia, scales are deeply important and will constitute the core of its definition, therefore of this book, and in some contexts place can be as significant than space, if not more so (Najib, 2020a).

Finally, space, scale and place involve shifting, stretching and mobile sets of relationships, processes and interactions (Massey, 1991). Here, I explain how an understanding of space in all its complexity depends upon an appreciation of social processes, and *vice versa*; how an understanding of social processes in all their complexity depends upon an appreciation of spatial forms. Adapting spatial form as an interpretive lens enables the study of Islamophobia constructions, experiences and negotiations to come to the fore. Meanings of space are key to a better understanding of Islamophobia, since this highly flexible phenomenon is situated, negotiated and contested differently through space.

Indeed, there are several types of Islamophobia. The different Islamophobic incidents show how Muslims or perceived Muslims can be excluded from certain spaces and places, thus from taking ownership of them. That is why it is important to include the spatial dimension in Islamophobia Studies and to explain how the concepts of Islamophobia and space are conjoined and clearly to determine the kind of land-use diffusion patterns that emerge (concentric, sectorial, residential, etc.). Besides, the nature of Islamophobia and the nature of space, and the relationship between them, relate and intersect to other important criteria that we need to take into account, such as gender, race, class and also political power, the functioning of the city system, and so on. This book will concretely establish that space, whatever form it takes (public, private, central, urban, etc.), is now recognized to be of major significance in shaping the geography of contemporary Islamophobia.

Islamophobia within Geography

The literature on Islamophobia is arguably too centred on critical treatments of racism. Not even within the discipline of Geography is the geographical component clearly theorized or explored empirically, despite an emerging literature. Within Geography, the term of Islamophobia began to be used in the late 2000s, notably in the subfield of social and cultural geography (Hopkins, 2019; Najib & Teeple Hopkins, 2020). Previously, Islamophobia was rather explored as a related topic, notably through the general study of Muslim identities and their marginalization in Muslim-minority countries (Dwyer, 1999; Naylor & Ryan, 2002; Peach & Gale, 2003; Falah & Nagel, 2005; Peach, 2006b; Phillips, 2006; Hopkins, 2007; Kwan, 2008; Hopkins & Gale, 2009; Mansson McGinty, 2012). Earlier geographical research on Muslims emerged first from studies on race and racism (Forrest & Dunn, 2011; Nelson & Dunn, 2017; Hopkins, 2019) and tended to focus on processes of racialization (Kobayashi & Peake, 2000) and identification through cultural representations (Dwyer, 1999; Falah & Nagel, 2005), as well as on residential segregation (Gale, 2007; Peach, 2006b; Phillips, 2006).

Similarly, in terms of books on Geography, there are several on Muslim identity and Muslim exclusion yet not explicitly on Islamophobia. For example, there is no book title mentioning the term and contributing directly to the geographies of Islamophobia. There are several influential volumes on the market that present the lived experiences of Muslims in specific spatial contexts. For example, the geographers Peter Hopkins and Richard Gale offer in *Muslims in Britain* (2009) a thorough and comprehensive account of what it means to be Muslim in today's

Britain. *Geographies of Muslim Identities* (2007) by Cara Aitchison, Peter Hopkins and Mei-Po Kwan illustrates the ways in which Muslim identities are constructed, represented, negotiated and contested in everyday life in a wide variety of international contexts. Richard Phillips, in *Muslim Spaces of Hope: Geographies of Possibility in Britain and the West* (2009), challenges what has come to be viewed as the 'Islamic problem' by asking what Muslims have to be hopeful about today and offering new ways of looking at social difference.

Given the rising interest in the 'Islamic problem' in the scientific geographical literature, the study of religious identities has increased significantly and, for some geographers, religion has even replaced race and ethnicity as the most significant in minority populations (Peach, 2006a, 2006b; Gale, 2013). I have always agreed with this analysis by Ceri Peach, seconded by Richard Gale, and I am happy to be able to acknowledge this here. There has indeed been a significant shift in discussing racism from race to religion. Peach (2006a) explained already in 2006 that the focus on racialized minorities mutated from 'colour' in the 1950s and 1960s (Banton, 1955; Rose, 1969), to 'race' in the 1960s, 1970s and 1980s (Rex & Moore, 1967; Smith, 1989), to 'ethnicity' in the 1990s (Bonnett & Carrington, 2000; Carrington et al., 2000) and to 'religion' in the present time (Runnymede Trust, 1997; Peach, 2006a; Gale, 2013; Najib & Finlay, 2020). Thus, in recent decades, there has been a significant growth in geographical studies that focus on discrimination in relation to religious identities, and especially to Muslim identities.

The geographical situations of Muslims in post-1990s studies are primarily shaped and expressed through the lens of oppression, especially related to deprivation, social exclusion and postcolonial conditions (Phillips R., 2009), and strongly contrast with older contributions of Muslim geographers of the 'Islamic golden age 8–14th centuries' (such as Al-Idrisi, Ibn-Battuta, Ibn-Khaldoun), a period when Muslim thinkers, travellers, traders, poets, philosophers, artists, scientists and engineers introduced new ideas, new technologies and new ways of understanding (Kong, 2009). Their findings did not spread throughout the Western world (or very little), and were eclipsed during the European colonization and Christian missions (ibid.).

Interest in religion in general, not only the Islamic faith, started to decrease with the beginning of world secularization situated by Wilson (1966) in the 1960s, in which religious thinking, institutions and practices were observed to have lost their social significance (Kong, 2009). This secular trend shaped the ways in which geographers approached the study of religion, with a greater focus on spiritualities (Bartolini et al., 2017). But the revival of religion in many places around the world and a growth in fundamentalism have given rise to a 'new' geography of religion, showing a shift from secular to post-secular status in which spirituality retains a significant role. More studies on the conflicts arising from secular representations of religious community have appeared. A range of themes is discussed by geographers, such as religious landscapes, religious identity formations, religious segregation, and so on.

This 'new' geography of religion started in the 1990s and 2000s and focused mostly on Muslim geographies (Najib & Finlay, 2020). We may wonder why this interest about Islam and Muslims is important to Geography: probably because it reflects the general tendency to focus on the margins instead of the dominant group, or because it classically reflects the Orientalist tendency when treating Islam and Muslims as objects of obsessive attention (Lewis, 2009). Mansson McGinty (2012) explains that scholars themselves can participate in the production of certain representations that frame Muslims as the 'Other', and we all need to be careful not to talk about Muslims and Islamophobia as an exception, as a unique racism, as the unique outcasts. This form of exceptionality would portray Muslims as an exception and, instead, we need to describe Muslims, particularly in the West, as full Western citizens affected by a form of anti-Muslim racism. This interest also responds to a range of pressing issues to do with Muslims in the West (Kong, 2009; Najib & Finlay, 2020) in line with the realities of public anxiety over Islamist extremism, urban disorders and terrorist attacks.

Studies on politics of religious space constitute a significant body of the geographical research on Muslim geographies, notably the study of mosques (Gale, 2009; Ehrkamp, 2007), schools (Hewer, 2001; Meer, 2007; Berglund, 2009; Dwyer & Shah, 2009) or *halal* food (Riaz & Chaudry, 2003; Isakjee & Carroll, 2019). Much of this work is from Britain and is focused on British concerns. Indeed, British studies have contributed to various pioneering perspectives and approaches, notably centred on the study of the body, building sites and neighbourhoods (Kong, 2009), and this needs to be spread to other regions of the world.

Finally, Muslim geographies in post-1990s studies have primarily been geographies of contestation, conflict and politics; that is why the transition to the study of geographies of Islamophobia was already framed in the context of negotiating the place granted to Muslims living in the West. Geographers researching religion started to contribute to Islamophobia Studies. Although not necessarily defining it, some began engaging with the term of Islamophobia (Dwyer et al., 2008; Mansson McGinty et al., 2012), while others started to define its contours in research in France, Sweden, Australia and the United States (Hancock, 2015; Najib, 2019; Teeple Hopkins, 2015; Listerborn, 2015; Dunn et al., 2007; Itaoui, 2016, 2020; Mansson McGinty, 2020). Following earlier literature, geographers also explored the process of racialization through violence and discrimination directed at visible signs of Islamic belonging (e.g. beards, headscarves, mosques) (Dunn et al., 2007; Hopkins, 2004). Drawing on theories of intersectionality, this racialized dimension of Islamophobia is also gendered, and feminist geographers have shown how the racist attacks and interpersonal aggressions in public spaces affect especially Muslim women who wear the *hijab* [6] (Hancock, 2015; Najib & Hopkins, 2019; Gökariksel & Secor, 2015; Listerborn, 2015; Teeple Hopkins, 2015).

In general, social and cultural geographers who work on Islamophobia usually 'overlook' its spatial dimension and continue to conceptualize this phenomenon simply as a form of systemic racism, just as sociologists, anthropologists and political scientists define it. This conceptualization is appropriate: sociologists and anthropologists mainly formulate concepts and definitions about Islamophobia and relate them to critical treatments of racism; and political scientists mostly analyse Islamophobia within existing power structures and examine anti-Muslim discourses. But geographers should be able to refine this definition by adding a spatial dimension and by arguing that Islamophobia is also a spatialized process that occurs at various scales (from the globe to the body (and mind)) and is connected between scales (Najib & Teeple Hopkins, 2020).

The advanced concept of Spatialized Islamophobia

The spatiality of Islamophobia needs to be better explained in future work, and this book is a first step. As previously explained, there are important academic publications on Islamophobia (Allen, 2010; Sayyid & Vakil, 2010; Massoumi et al., 2017; Morgan & Poynting, 2012; Bayrakli & Hafez, 2019), yet none explores its spatialization in a complete way, even in the discipline of Geography (Dwyer et al., 2008; Mansson McGinty et al., 2012; Hancock, 2015; Gökariksel & Secor, 2015; Listerborn, 2015; Hopkins, 2019). Building on this geographical tradition of 'overlooking' the spatial dimension of Islamophobia, I co-edited with my colleague Carmen Teeple Hopkins a special issue of *Social and Cultural Geography* entitled 'Geographies of Islamophobia' in 2020, where we introduced the spatialized process of Islamophobia. The special issue includes six distinct contributions from both early career and established scholars, all of whose research substantively engages with Islamophobia in social and cultural geography (Anna Mansson McGinty, Claire Hancock, Rhonda Itaoui, Lauren Fritzsche, Lise Nelson, Peter Hopkins and Robin Finlay), and we bring together a cohesive understanding of the geographies of Islamophobia. After this publication, I personally felt the need to develop the advanced concept of Spatialized Islamophobia in greater depth and to offer additional insights to both complement Islamophobia Studies and enhance this type of research in Geography.

This book insists that Spatialized Islamophobia is multi-scalar. It operates at various spatial scales from the global to the micro-local. Islamophobia affects many spheres of our society, and it can be seen as a matter of social justice to regard it as contingent on the spatialized process operating in space as a whole. The spatialized process means that space is understood as a relationship between objects that exists only because those objects exist and relate to each other (Harvey, 1973). Here, the notion of 'spatialized' is broad enough to encompass the notions of space and its related notions, such as scale, place, geography, global, local, and so on, as they are relevant to the contexts in which Islamophobia is forged. Geographers are particularly interested in working at the level of aggregation to apprehend any territorial organization of societies. And it is precisely the relationships between social and spatial forms and between different aggregate levels that describe the process of Spatialized Islamophobia. Its definition is provided in bold in Box 1.1 for ease of reference. The concept of Islamophobia is already much discussed (and sometimes contested). Therefore, adding to existing definitions can be a challenging task, and here I only develop and refine this new concept of Spatialized Islamophobia from a previous study (Najib & Teeple Hopkins, 2020).

Box 1.1 Definition of Spatialized Islamophobia

Spatialized Islamophobia is a spatially pervasive form of anti-Muslim racism that occurs at various interrelated spatial scales (globe, nation, urban, neighbourhood and body (and mind)) and whose contours, effects, intensity and functioning vary accordingly.

This definition of Spatialized Islamophobia, which is amply and implicitly reflected throughout this book, is based on the literature that shows that Islamophobia is: 1) a systemic anti-Muslim racism that targets markers of Muslimness or perceived Muslimness (Halliday, 2003; Naber, 2008; Garner & Selod, 2015; Allen, 2010; Sayyid & Vakil, 2010; Scott, 2007; Massoumi et al., 2017); and 2) a spatialized process that occurs at various interrelated spatial scales (Najib & Teeple Hopkins, 2020). Spatialized Islamophobia shows that Islamophobia is everywhere and affects all spatial scales. Even more important are the interrelationships between the global scale, the national scale, the urban scale, the infra-urban scale and the scale of the body (and mind) in the study of Spatialized Islamophobia than detailing each scale one by one. It is important to pay attention to these interrelationships to better understand how this anti-Muslim racism manifests itself and moves from one scale to another. Spatialized Islamophobia is influenced by global representations, national policies, urban mobility strategies and emotional behaviours that are in constant communication. Clearly, the various scales matter, and their interrelationship requires the special section below that considers it as a significant influence on how Spatialized Islamophobia is experienced and practised.

The interrelationship between the scales of Spatialized Islamophobia

According to the scale under study, Islamophobia changes its contours, its effects, its intensity and its functioning, so that it is difficult to believe that it is the same phenomenon and that the scales are separate. Yet Spatialized Islamophobia is indeed a single phenomenon, and the scales are indeed distinct from each other but interrelate with each other. It is therefore important not to isolate each scale to better understand this phenomenon as a whole, and to focus on their interrelationship. Here, I speak of 'interrelationship', but it is perhaps more accurate to speak of 'entanglement' or 'intertwining'. In fact, these scales (globe, nation, urban, neighbourhood and body) interlock: the individual exists in his/her residential space, included in the urban space, included in the national space, included in the global space. Therefore, none can be understood in isolation since each interacts with others and is defined with respect to the role it plays vis-à-vis the others.

That said, in order to make the reading easier, the next chapters will set out more in detail the salient features of the three main levels of scales: 1) global and national Islamophobia; 2) urban and infra-urban Islamophobia; and 3) embodied and emotional Islamophobia. It is important to understand in brief these scales that are studied:

1   Islamophobia is a global phenomenon (Morgan & Poynting, 2012; Bayrakli & Hafez, 2019; Osman, 2017) that can be observed in various national contexts, especially in the general stigmatization of Islam and Muslims emanating from their misrepresentation (Gale & Hopkins, 2009) in the media and politics after significant events in the international arena. To be more accurate, terrorist attacks perpetrated in the name of Islam have fuelled anti-Muslim sentiment around the world, especially in the West, leading to more open racism against Muslims or people identified as such. Anti-Muslim political discourses mainly question the compatibility of Islam with Western values, and some laws attempt to reinforce the idea that there is no place for visible signs of Islam. The political spaces of anti-Muslim sentiments and crimes differ between macroregions and countries, showing the important role of policies and laws in religious diversity and secularism.

2   The spatial elements of Islamophobia that display its operation and distribution are more readily appreciated in the city. The map is an important tool in Geography, and here geographers can play a significant role in measuring and mapping the phenomenon in order to appreciate fully where it happens, how it functions and how it can be challenged. The mapping of Islamophobia reveals a specific logic of spatial distribution, showing the importance of particular spaces and places as well as processes of exclusion/inclusion better observed at the infra-urban scale (the scale of districts or boroughs). These geographical tensions undermine the spatial mobility and belonging of victims, especially veiled Muslim women. They mainly experience anti-Muslim practices in everyday encounters, and this daily violence implicitly questions their right to certain areas of the city (Lefebvre, 1996) and points to their strong spatial attachment to their home neighbourhood.

3   Islamophobia is both embodied and emotional (Mansson McGinty, 2020), and it has impacted strongly on the everyday lives of Muslim bodies and minds. Veiled Muslim women are particularly affected, due to their great visibility. For these women, wearing the *hijab,* for example, is a daily challenge (Bowen, 2007; Scott, 2007; Fernando, 2009), but all Muslim bodies (bearded bodies, full-covered bodies, veiled bodies, etc.) are concerned with embodied and emotional Islamophobia. Their experiences of oppression have led them to develop and adopt new behaviours, especially strategies of invisibility and normalcy, in response to Islamophobia. These oppressions have also pushed some Muslim bodies towards more private space, such as their homes. Yet the intimate space of the home can also be a place where Islamophobia manifests itself. Family members who have differing religions or interpretations of Islam may develop Islamophobic attitudes to visible Muslim relatives who often try to resist in a non-confrontational way.

Detailing the three main levels of scales has already shown how spatialized and multi-scalar Islamophobia is and how the various scales interrelate, intersect and reflect on each other in significant ways. Considering only one scale can affect our general observations and findings, since the well-known problem of spatial aggregation (or Modifiable Area Unit Problem (MAUP)) shows that a just distribution across a set of territories defined at a single scale does not necessarily mean a just distribution at another, or among individuals (Harvey, 1973; Openshaw & Taylor, 1979). That is why it is better to emphasize the various multiscalar findings of Islamophobia, from global to intimate.

Taking into account the scales of Spatialized Islamophobia all at once can uncover important aspects of the phenomenon, such as different forms of racism, different ways of governing, different political boundaries between Muslims and non-Muslims, and also different socio-spatial effects of Islamophobia, different experiences of public space, different spaces of belonging, and so on. Spatialized Islamophobia changes its functioning according to the scale under study: it is as if everything changes from one spatial scale to another. The study of their interrelationships indicates that Islamophobia projects a reality from the global scale to the local scale. Islamophobia is therefore a glocal process that 'glocalizes' global hatred sentiments by impacting on the everyday lives of local Muslims living in a specific context. This glocalization particularly raises awareness about the local repercussions of global anti-Muslim racism and their related national policies. Although the term was coined by the Japanese business world (Robertson, 1995), it can be applied to other fields, notably the social sciences, where the glocal approach has already shown the importance in everyday lives of connecting the global scale with the micro-local scale, especially in matters related to fear and inequality (Pain & Smith, 2008; Pain, 2009; Pain & Staeheli, 2014). Any phenomenon with global and local aspects can be discussed using this glocal approach, and particularly Islamophobia (Morgan & Poynting, 2012). Indeed, micro-local, local, national and global contexts emphasize how differently Islamophobia is practised on an everyday basis.

Figure 1.1 presents the funnel process of Spatialized Islamophobia, yet it does not explain that this is only a one-way process flowing from the global fear of Islam into people's everyday lives. It only traces the interrelationships between the various scales of Spatialized Islamophobia that emerge through the chapters in this book. Scales are not to be seen as simply distinct entities but as relations among one to another. To be more accurate, the global Islamophobia spread by the media has local consequences (sometimes dramatic), which in turn can question democratic values and challenge the entire world. Islamophobic experiences lived at the scale of the individual or the scale of the body, notably a woman's body, can clearly test our liberal democracies. The resistance emanating from Islamophobic acts targeting veiled Muslim women, for example, can challenge worldwide democratic values such as freedom, diversity, feminism, and so on.

Spatialized Islamophobia is both a top-down and bottom-up process, as illustrated by the two-way arrows in Figure 1.1. On the one hand, the top-down process means that global representations of sensational events spread by the media can lead to the implementation of specific national laws that aim to reduce the Muslim presence and control Muslim visibility. These discourses and policies can foster the rise of concrete Islamophobic acts on the urban ground. The acts of violence may occur in specific spaces and not necessarily in others, drawing a real

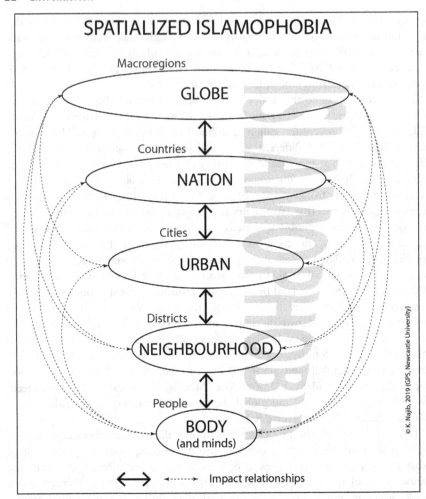

Figure 1.1 The glocal process of Spatialized Islamophobia

geography of Islamophobia. The Islamophobic climate in certain boroughs and places may lead victims to develop strategies to avert potential situations of violence or discrimination. Their fear and anxiety can be mentally mapped to specific spaces to trace an emotional geography of perceived risk.

On the other hand, the bottom-up process may show that victims of Islamophobia perceive space emotionally, and this embodied Islamophobia can lead them to use mainly safe spaces and to adopt specific behaviours to ensure their security. They are pushed into marginal and private spaces, avoiding the rest of the city. Their partial use of the city excludes them from the dominant and privileged centres, reflecting Muslims' national marginalization and global stigmatization. The link between the global and the local and between the local and the global is not always clear-cut, but this first attempt at explanation, schematized in Figure 1.1, shows the interrelationships (the two-way arrows) between the scales, which are directly and successively interlocked. Obviously, there are further interrelationships between the other scales, illustrated by dotted two-way arrows in Figure 1.1, just as there are more appropriate verbs to explain the various impact relationships.

Spatialized Islamophobia must be considered both between and within scales. Indeed, it differs both between scales and at a single scale. For example, at the national level and according to the cultural and political differences, Islamophobia may be triggered in one country by State laws and in another by the people. The glocal approach advocates a specific integration of elements from global to local paradigms that may, for example, maximize cross-cultural comparability between countries, city size effects between cities, social differences between boroughs, or physical and emotional vulnerabilities between people. Spatialized Islamophobia appears therefore as a complex system composed of dynamic relationships defined by the spatial context in which it happens, in line with its political, cultural and social characteristics.

The globalization and the localization of Islamophobia cannot be separated into distinct aspects, and a shift has been noted from global representations of Islam and its followers to the real impacts on Muslims' everyday lives. This shift has even been noticed in the literature: for example, the Runnymede Trust's first report on Islamophobia in 1997 defined it as 'an unfounded hostility towards Islam' (1997: 4) that leads to a fear or dislike of Muslims; while its second report, released in 2017, defined it as an anti-Muslim racism, including 'any distinction, exclusion or restriction against Muslims' (2017: 1). This glocal reading should nevertheless state an appropriate balance between the important global negative representations of Islam and Muslims and the local realities on the ground, where the great majority of Muslims live on good terms with their neighbours, colleagues and friends. I am not saying that local Islamophobia is less important; the rise of Islamophobic violence is alarming and affects Muslims on a daily basis. I am just saying that when a Muslim is portrayed by the media, in the vast majority of cases, it is a negative representation (Saeed, 2007; Allen, 2010), while we witness most Muslims living peacefully in their local area. This raises the difficulty of reading Islamophobia when it is not physically visible; someone with Islamophobic sentiments will not necessarily demonstrate them openly, in a certain context.

The glocal process of Islamophobia highlights several specificities. According to the space under study, the nature of the Islamophobic incident is different, just as the place where it happens is different, just as the victims and the perpetrators are also different. Indeed, Islamophobic acts differ from one scale to another: they can be racist discourses and laws, institutional discrimination, racial profiling, verbal abuses and physical aggressions of people, or profanation and degradation of religious buildings. They can take place in public institutions, in the workplace, at the airport, in public areas or on public transport. They can affect men, women, children, Muslims or non-Muslims.

Indeed, the victims differ from one scale to another. For example, global Islamophobia in line with the global threat of terrorism and global conflation between

Islam and terrorism mainly concerns Muslim men. National Islamophobia, linked to laws on religious dress, mainly concerns visibly Muslim women. Women's bodies represent the focus of attention most of the time, whether with regard to the *niqab* or the simple *hijab*. Therefore, we can wonder whether veiled Muslim women are the most targeted because they are more fragile and vulnerable in public areas or, rather, because dominant narratives and laws portray them as the 'enemy within' of Western values. It is as if a glocal identity is projected onto them; meanwhile, at the personal level, they can be totally uninterested in politics and public debates.

Finally, this glocal approach implies different research methods. When studying a particular spatial scale, methodological constraints are imposed upon us. Suitable methodological decisions need to be made with a clear understanding of the phenomenon of Spatialized Islamophobia. Some aspects are identifiable through quantitative measures, while others are observable only through more qualitative approaches. For example, if we want to quantify the phenomenon of Islamophobia in various countries or map its spatial distribution in a given city, quantitative data are needed. But if we want to know how Muslims move across the city or how they behave, then more qualitative material such as individual interviews are needed.

The various scales that explain the phenomenon of Spatialized Islamophobia are frequently submerged in the content of existing literature, but they tend to emerge at crucial points for explicit consideration as signposts are sought to guide the analysis. They are rarely considered all at once, and Table 1.1 details all the scales of Spatialized Islamophobia. In this chapter, I try to show the effects of certain 'hidden mechanisms' that tend to be obscured by our difficulty in considering Islamophobia within a spatial whole and using different methods. Exploring Islamophobia from various spatial perspectives necessarily implies the use of both quantitative and qualitative approaches in order to understand not only the general context of this phenomenon but also its details. This can represent a breakthrough, in Islamophobia Studies, which is analysed and criticized better in the fifth and final chapter. But before, I analyse the three main levels of scales in the following chapters: 1) global and national Islamophobia (Chapter 2); 2) urban and infra-urban Islamophobia (Chapter 3); and, finally, 3) embodied and emotional Islamophobia (Chapter 4).

Conclusion

Spatialized Islamophobia is a new concept that I have developed for three reasons. First, as a geographer interested in Islamophobia Studies, I find it consistent that Islamophobia is a spatialized process, and therefore entirely possible to analyse it with a geographical lens.

Second, writing up the SAMA research project conducted in a British university allowed me to move from a first geographical impression to real, in-depth exploratory research. Indeed, the SAMA project considers the impact of Islamophobic incidents on space and people, using a mixed-methods approach based on quantitative data

Table 1.1 Methodological issues of Spatialized Islamophobia

Spatialized Islamophobia	Methods and objectives	Spaces concerned
Global Islamophobia	Historical and quantitative approach – methods centred on global representations: • **Historical and political analysis:** Critics of the power structures and the dominant groups that produce 'the Muslim problem'; • **Sociological literature:** Focus on Islam and Muslims as a monolithic bloc; • **Globalization of Islamophobia:** Muslims around the world and their public and political representations; • **Comparative approach**: Comparison between different macroregions, different global statistics, etc.	Muslim-minority / Political / Power / Dominant / Media spaces
National Islamophobia	Political and quantitative approach – methods centred on national discourses and policies: • **Political analysis:** Policies and laws targeting expressions of Muslimness; • **National belonging:** The visibility of Muslims and their citizenship and belonging; • **Comparative governance:** Cross-national analysis according to the different cultures, politics, immigrations, statistics, etc.	International / Political / Power / Legal / Institutional / Visible / Secular / Public spaces
Urban Islamophobia	Geographical and quantitative approach – group-based methods: • **Quantitative data collection**: Analysis of statistical data collected from the police, NGOs, etc.; • **Mapping**: Spatial distribution of quantitative Islamophobic acts; • **Graphical modelling**: Synthetical schemes of distribution logics and spatial patterns (urban models); • **Statistical and cartographic comparison**: Cross-urban analysis of the extent of the phenomenon between different cities and within the same city. Comparison with other geographies (social, economic, ethnic, religious, etc.).	Urban / Public / Geographical / Social / Racial / Stigmatized spaces
Infra-urban Islamophobia	Geographical and mixed-methods approach – methods based on groups and individuals: • **Qualitative (and quantitative) data collection:** Analysis of qualitative data directly collected from the victims (but also quantitative data collected from existing survey, the police, NGOs, etc.); • **Mapping**: Mental maps on perceived Islamophobia; • **Spatial analysis**: Identification and negotiations of spatial boundaries (exclusion/inclusion, safe/unsafe, etc.); • **Spatial mobility**: Analysis of victims' spatial strategies (avoided spaces, used spaces, etc.); • **Geographical anchorage:** Feelings of spatial belonging and the importance of the home neighbourhood for the victims; • **Cartographic comparison**: Comparison between 'real' and 'perceived' geographies of Islamophobia.	Residential / Everyday / Safe / Perceived / Exclusive / Marginal spaces

Spatialized Islamophobia	Methods and objectives	Spaces concerned
Embodied and Emotional Islamophobia	Qualitative approach – individual-based methods: • **Qualitative data collection:** Central to this study is the analysis of individual interviews with victims; • **Behavioural strategies of Muslim bodies (and minds):** Performing invisibility (by hiding their Muslimness) and normalcy (by displaying that they are good integrated Muslims) through the analysis of their experiences, concerns, fears, emotions, moods, etc.; • **The home space:** Performing intimacy (securing their bodies and minds by staying at home as often as possible) through the same analysis; • **Intimate Islamophobia:** Islamophobia within the family (parents, in-laws, etc.).	Emotional / Private / Intimate / Home / Family / Cultural / Psychological spaces

Note: Sometimes only one type of space is mentioned in the spaces concerned column, yet the opposite is implied (e.g. when mentioning safe spaces, unsafe spaces are included).

(collected from community associations and the Metropolitan Police) and qualitative data (directly from victims of Islamophobia on the databases of NGOs fighting against Islamophobia). Specifically, this project focuses on French and British Islamophobia (that also transfers to a broader international context) and analyses the spatial distribution of Islamophobic acts, the spatial and behavioural strategies of victims, their spatial scales of belonging, and the interrelations with other types of discrimination such as sexism, racism, ageism and classism.

Third, the special issue, Geographies of Islamophobia, introduces the spatial dimension of Islamophobia leading to the development of an important area of research in human geography. Such research is attentive to the intersections of religious identities and the role of religious abuses in the formation of communities and sense of belonging. All these reasons – and also particularly my personal identity and experience – strongly encouraged me to develop further, through this book, the concept of Spatialized Islamophobia, showing how it occurs at various interrelated scales (ranging from the globe to the body).

Finally, a spatialized and multi-scalar reading of Islamophobia primarily indicates the importance of spatial interrelationships. Other relationships are also significant in the study of Spatialized Islamophobia, in particular power relations. These are apprehended at the largest scales. They are detailed in the next chapter, which explores this pressing real-world problem of global and national Islamophobia.

Notes

1 The SAMA project was written by Dr Kawtar Najib under the supervision of Professor Peter Hopkins, supported by the European Commission through the H2020 Marie Sklodowska-Curie Actions [Horizon 2020-MSCA-IF-2015–703328].

2 The special *SCG* issue, 'Geographies of Islamophobia', was co-edited in 2020 by Dr Kawtar Najib and Dr Carmen Teeple Hopkins.

3 When I talk about racialized researchers here, I mean racially minoritized researchers as 'White' is also a race.
4 *Laïcardes* organizations means those that use *laïcité* (French secular values) to combat religion and erase religious discussion from society.
5 The *laïcité* guarantees the neutrality of the State, the French administration and their representatives. Indeed, the *laïcité* fosters the free exercise of all religions and has never excluded any religion, protecting minority religious groups (Baubérot, 2012). Finally, citizens and users of public institutions have the right to express a belief or non-belief when there is no disturbance of public order.
6 The *hijab* is a veil that some Muslim women wear in order to cover their hair, neck, ears and chest. They wear it in the presence of men outside of their immediate family (their father, brother, son, uncle, nephew, husband and father-in-law can see them with hair uncovered).

2　Global and national Islamophobia

Introduction

Islamophobia is a global phenomenon (Morgan & Poynting, 2012; Bayrakli & Hafez, 2019; Osman, 2017) that may be observed in many areas of the world. This is especially the case due to the contemporary international misrepresentation of Islam and Muslims in the media and politics following major terrorist attacks. The 9/11 and subsequent attacks, as well as the so-called 'war on terror', have fuelled an anti-Muslim climate around the world, leading to a more open racism against Muslims. In the context of globalization and new communication technologies, the spread of anti-Muslim sentiment is impressive and has now reached many of the most diverse countries, whether located in the West or elsewhere, and whether it involves Muslim-minority contexts or not. Indeed, Spatialized Islamophobia differs by context and may also be experienced in predominantly Muslim societies, notably in countries such as Turkey, Tunisia and Egypt, for various socioeconomic and political reasons. In predominantly non-Muslim societies, Islamophobia follows other processes, in particular domination by the 'majority' over the 'minority'. It can reach significant levels of violence, whether in non-Western countries like in Asia (for example in Myanmar, China and India) or in the West (such as in Europe, the United States and Australia).

Islamophobia in the West represents the study area that is most discussed (Allen, 2010; Beydoun, 2018; Sayyid & Vakil, 2010; Bayrakli & Hafez, 2019): even if Islamophobia is not restricted to the geography of the West, it is still important to explore this context while emphasizing its spatialized nature. The spatialization of Islamophobia exposes a strong relationship between the 'visibility' and 'invisibility' of markers of Muslimness, and it is particularly important here to understand why the visibility of Muslim populations has become a serious concern in Western countries that are supposed to be liberal democracies. Muslim populations living in the West, as well as their experiences, conditions and claims, challenge such democracies and their policies strongly. National Islamophobia in the West questions the compatibility of Islam and Muslims with Western values, and some anti-Muslim laws use important values such as secularism, freedom, modernity and feminism to reinforce the idea that there is no place for visible signs of Islam (e.g. the 2004 headscarf ban in public schools in France, the 2009 ban of

DOI: 10.4324/9781003019428-2

minarets[1] on mosques in Switzerland and the 2017 'Muslim ban' on immigrants and refugees to the United States).

At a national scale, countries have issues between governance and faith to varying extents. There are often differing interpretations of religious diversity across countries, and the specificity of national histories is important to an understanding of how democratic concepts and racial considerations become manipulated to justify how Islamophobia becomes legally enshrined in law (Allen, 2010; Modood, 2009; Teeple Hopkins, 2015; Najib & Teeple Hopkins, 2020). After analysing the processes of Spatialized Islamophobia around the world and more particularly in the West, the last section of this chapter focuses on the French and British examples, in line with the SAMA research project. Islamophobia in France and the United Kingdom is well documented (Hancock, 2015; Najib & Hopkins, 2020; Najib, 2019; Dwyer, 1999; Allen, 2010; Modood, 2009; Hopkins & Gale, 2009), but this section offers a cross-national comparison, contrasting the French Republican model with the British multicultural political model. The political spaces of anti-Muslim sentiments and the important role of policies and laws on religious diversity will reveal the actual place granted to Muslim populations in these two countries.

Muslims and global Islamophobia

Spatialized Islamophobia manifests itself differently across the world, and is globalized in the sense that it does not spare any continents or macroregions. It is at a varying stage in each country, depending on its geopolitical interests. It seems at an early stage in some, while it poses a serious threat to life in others. More accurately, Islamophobia can be read as a contempt of religious Muslims by a certain elite class in predominantly Muslim countries (Bayrakli & Hafez, 2019) or as an anti-Muslim stigmatization in Western countries (Morgan & Poynting, 2012), or even as a crime against humanity in specific countries located in Asia (Osman, 2017, 2019; Majeed, 2019). The media bears significant responsibility for global Islamophobia (Ameli, 2004; Githens-Mazer & Lambert, 2010; Saeed, 2007), and therefore, whatever the spatial context, Islamophobia follows a world-system that depicts Muslims as: 1) a monolithic bloc (Halliday, 2003); 2) terrorists (Kundnani, 2014; Hopkins, 2008; Staeheli & Nagel, 2008); and 3) culturally backward and opposed to modern civilization (Said, 1978; Kumar, 2012; Bayrakli & Hafez, 2019). But to be able to make such conclusions, it is essential to know what the Islamic faith is and how it is practised throughout the world.

First, far from being a homogenous community, Muslims around the world represent a wide variety of ethnicities, religious interpretations and practices, and linguistic and cultural backgrounds, proving their great heterogeneity and conflicting identities. Muslims are regarded as a group with no internal diversity, variation or contradiction, yet they are exceptionally diverse around the globe in terms of social, cultural and political organizations. According to the Pew Research Centre, in 2015 Muslims made up 24% of the world population, represented by 1.8 billion individuals (Lipka, 2017). Therefore, the plurality of the various Muslim populations cannot be denied: Muslims can be Sunnis, Shiites, Arabs, Asians, Black Africans, Whites, from privileged classes or disadvantaged families, and so on.

Islam as a faith is the following Abrahamic faith after Judaïsm and Christianism, and believes in all the prophets and messengers of God. It is a monotheistic religion with specific worship and practices, exactly like other religions. There are two major denominations of Islam (Sunni and Shia), and within Sunni Islam there are four main schools of Islamic jurisprudence (Maliki, Hanafi, Shafii and Hanbali), while there are three major divisions within Shia Islam (Ismaili, Jafri and Zaydi). There are also other movements, such as Sufism, Ahmadiyya, and so on. This brief presentation of what Islam is demonstrates the internal diversity of Islamic cultures and therefore contradicts the idea of homogenization of the Muslim populations. In this sense, I prefer to use the designation 'Muslim populations', in the plural, to attest to their diversity, even if I sometimes use the simplistic label of 'Muslims'.

Second, discourses about the responsibilities of Muslims in the wake of the various terrorist attacks imply that the Islamic identity and faith are contingent on terrorism (Kwan, 2008; Runnymede Trust, 2017; Cohen & Tufail, 2017; El Zahed, 2019). But Islamist terrorism primarily affects Muslim countries; therefore, global terrorism cannot be due to Islam. Muslim populations are globally depicted as terrorists, especially due to the flashpoint of the 9/11 attacks on New York's World Trade Centre. Since then, markers of Muslimness have become synonyms of radicalization resulting in exclusion of Muslim populations, especially practising ones. Instead of differentiating terrorists from civilians (Mamdani, 2004), religious Muslims are considered 'bad Muslims', as opposed to 'good moderate Muslims'.

This process of exclusion means that many Muslims become in their everyday lives the victims of discrimination, harassment, racial and religious profiling, or verbal and physical assaults (Peek, 2003; Beydoun, 2018), and this occurs in several countries. Markers of Islamic piety have nothing to do with extremism, because terrorist attacks have targeted religious Muslims themselves, as in the rarely reported attacks in Iraq, Somalia or Pakistan. This simplistic conflation of Islam with terrorism rather refers to the existing global biases and stereotypes coming from Islamophobic political and media narratives (Cohen & Tufail, 2017). Consequently, these narratives that depict Muslims as potential terrorists and oppressors make it difficult to see them as a disadvantaged minority victim of religious hatred themselves (Meer & Modood, 2019).

Third, it cannot be that a quarter of the world's population (Lipka, 2017) follows a backward religion. This view of backwardness mostly comes from Western imperialism, which equates religious considerations in general and Islamic traditions in particular with dogmatism, as opposed to modernism and science. The important intellectual contributions of Muslim scholars over several centuries in areas such as Algebra, Astronomy, Geography, Physics, Mathematics and Medicine should not be so quickly forgotten (El Zahed, 2019); nor should their advocacy of social justice (Abdelkader, 2000; Mehmet, 1997). Instead, we have to ask ourselves how colonization overshadowed these findings (Kong, 2009). The common understanding is that the West is seen as positive, advanced and democratic, with Islam being representative of ignorance, intolerance, patriarchy and oppression

(Hopkins, 2009). This binary vision particularly targets visibly religious Muslims and illustrates how religious Muslims and non-religious Muslims represent conflicting views about the Muslim identity and how they respectively signify the opposition between traditional indigenous values (i.e. the backward religious Muslim) and modern Western values (i.e. the Republican and secular ideal citizen) (Ahmed, 1992; Gökariksel & Mitchell, 2005).

The diversity of representations of the Islamic faith and their followers in various parts of the world mirrors the complexity of Islamophobia. Around the globe, Spatialized Islamophobia can be analysed through the domination relationship between the majority and the minority. On the one hand, in Muslim-majority contexts, Islamophobia is witnessed in public debate. But the most striking element I take from Bayrakli & Hafez's (2019) collection of writings on Islamophobia in Muslim-majority societies is the hegemony of Western values adopted by many Muslim elites living in these countries who criticize, despise and undermine the traditional Muslim identity of the masses. In this context, Islamophobia depicts religious Muslims as a threat to the modern secular way of life, as observed in Western discourse. This desire to attain Western secular modernism in Muslim societies refers to Western Orientalist and self-Orientalist processes that relocate the figure of the subaltern Muslim subject in the world (Grosfoguel, 2011; Sayyid, 2014). Faced with this identity crisis, the Muslim elites disconnect from their own Islamic tradition, reaffirm their superiority over conservative Muslims (seen as uncivilized and backward) and problematize Islam in the same way as was the Church in European history (Bazian, 2019). Islamophobia in Muslim countries rather describes the power relation between the powerful and the powerless; that is, between Westernized secular Muslim elites and the conservative Muslim masses (Bayrakli & Hafez, 2019). This divide is complex: it finds its source in Orientalism (Said, 1978) and refers to dissimilar educations, generations and states of mind. Conservative Muslims can suffer from various forms of Islamophobia, implying not only rejection by specific schools and jobs, and exclusion from modern restaurants, hotels and public events (such as concerts, for example), but also the preference for Western clothes, the spread of Islamophobic comments through the media and the development of secular policies, such as the headscarf ban (for example in Turkey and Tunisia). Having said that, Muslim elites and their secular-Westernized vision are still not in power in these countries and do not constitute the dominant social group in terms of either demography, history or belonging, unlike conservative Muslims. Therefore, in my opinion, the relationship between the powerful and the powerless that Islamophobia describes in Muslim societies should not be seen as a process requiring total domination, unlike Islamophobia in Muslim-minority countries in which Muslims are part of neither the demographically dominant majority nor the powerful elite.

In Muslim-minority contexts, on the other hand, Islamophobia describes the clear domination by the majority over the minority. The minority status of Muslim religious groups can create situations of marginalization and exclusion and lead to experiences of discrimination and violence (just like other minority groups who face anti-Jewish, anti-Black or anti-LGBT hate crimes, for example). But it is important to distinguish those Muslim-minority contexts in Western countries (represented as models of democracy) from those in non-Western countries, particularly those in Asia where Islamophobia can be extremely strong (Osman, 2019). In some Asian countries, it is more alarming than the consistent anti-Muslim rhetoric in political and public debates in Western societies. It has serious implications for the life of Muslims, since it can lead to the literal death of entire groups of people simply because they are Muslim. Their humanity is denied, since they are seen and imagined as de-individualized objects within a collective construction. This vision is ongoing in our contemporary world, and in such countries, Islamophobia produces systemic racism that goes so far as to kill hundreds of thousands of Muslims just because of their faith. This is the case with the Rohingya community in Myanmar, killed at the hands of Myanmar's Buddhist military or corralled into detention camps (Ibrahim, 2016). Villages have been burned to the ground and many Rohingyas have been forced to flee to Bangladesh (Merelli, 2019). Similarly, the Uighur Muslim minority in China is currently suffering major abuses just because they are Muslims; the vast majority are locked up in detention camps, theoretically to be re-educated against extremism. In fact, they are tortured into renouncing their culture and religion (Maizland, 2019). In India, Muslims also experience alarming Islamophobic violence. This worsened in early 2020 due to the new Citizenship Amendment Act, which questions Muslims' citizenship (Saikia, 2020). Narendra Modi's government requires Muslims to prove their Indian identity, because their Islamic faith renders it suspect (Iftikhar, 2020). Muslims' houses have been torched and Muslims have been killed in connection with Modi's rhetoric.

That said, it is important to remember that this alarming level of Islamophobia is not unique to these regions of Asia. Islamophobia reached a level comparable to the greatest crimes against humanity in Europe not so long ago, in the 1990s. In July 1995, the genocide of Srebrenica in Eastern Europe led to ethnic cleansing of Bosnian Muslims in Bosnia-Herzegovina. In the current climate, however, Islamophobia in Europe and more broadly in Western society does not reach this level of crime. Violence and abuses are not negligible, but take the form of isolated attacks upon Muslims or perceived Muslims.

Nonetheless, a worrying new phenomenon has appeared in recent years: the murder of groups of Muslim worshippers in their local mosques, such as in January 2017 in Quebec City in Canada or in March 2019 in Christchurch in New Zealand (Loewen, 2019), involving far-right terrorist groups including White supremacists known to be avid readers of Islamophobic thought. In Western countries, Islamophobic thought mainly focuses on the growing visibility of Muslims in the public space and makes it problematic. Islamophobia in the West particularly targets Muslims because of their racial attributes, unlike Islamophobia in Muslim societies or in Asia, where Muslims and non-Muslims (or secular Muslims) share the same ethnicity. In the West, it describes a process of racialization and Othering; that is why it is particularly important to continue to study Spatialized Islamophobia in the West.

Muslims and Spatialized Islamophobia in the West

The construction of the 'Muslim problem'

Religious belonging has once again become an important issue in contemporary societies and politics, and in the wake of global changes the Islamic faith has become the new target, especially in the West (with the migration crisis, resistance movements against poverty, the return of isolationist and nationalist politics, the growth of racist groups, etc.). This interest in Muslimness has been illustrated over the past thirty years in numerous geopolitical events in which Islam has been central. Examples are the terrorism carried out in the name of this religion (9/11 and Paris attacks), debates and policies about religious freedoms in Europe (the headscarf ban in French schools), anti-immigrant protests focused on Islam (the 'Islamization' of Europe protests in Germany) and the reduction in the number of Muslim immigrants and refugees (former American President Donald Trump's 'Muslim ban' from seven Muslim-majority countries), to name just a few.

The politicization of the Islamic faith and the visibility of Muslim populations living in the West have become problematic. Muslims represent a growing visible minority in many Western countries, yet are still a small minority. In 2016, they represented nearly 5% of the total population in Europe (Hackett, 2017), 1.1% in the United States (Lipka, 2017) and 2.6% in Australia (Farrugia, 2019), according to the Pew Research Centre and the Australian population census. More accurately, in Europe the number of Muslims rose from 19.5 million in mid-2010 to 25.8 million in mid-2016 (Hackett, 2017), making them the largest single minority religious group, especially in France and Sweden where the proportion is higher (8.8% and 8.1%, respectively, of the total population). Both the growing share of Muslim populations living in the West and the visibility of their religiosity (such as mosques, headscarves and beards) give rise to fears, resulting in a range of Islamophobic occurrences from anti-Muslim policies to physical attacks perpetrated directly in the public space.

Islam, in all its institutional, spiritual, cultural and social variations, is at the heart of public debates in which it is argued that the Islamic religion is incompatible with Western values (Halliday, 2003; Gest, 2010). Muslims living in the West have been mainly framed as foreigners who do not belong in Western countries. These stereotypes are problematic because they represent 'Islam' and the 'West' as two antithetical categories, denying the plural identity of millions of Muslim and Western citizens. Many debates about 'Islam and the West', 'Islam and democracy', 'Islam and security', 'Islam and secularism' and 'Islam and postmodernity' (Gale & Hopkins, 2009) generally represent Islam as a problem and thus tend to frame Muslims' integration as a challenge (Modood, 2008), as if they were the only minority religious group with difficulty in integrating into society. Muslims are seen as 'invaders' of the West, and this crude binary between Muslims and non-Muslims echo the binary between non-Whites and Whites, revealing forms of domination over the former colonized by the former colonizer (Manji, 2004; Hopkins & Smith, 2008). This White imperialism establishes a malevolent new form of the 'clash of civilizations' thesis (Maddox, 2005), reinforced by the fact that Western countries are engaged in wars against certain Muslim-majority countries.

Kumar (2012) argues that colonial domination has continued to racialize Muslims as a political tool in order to maintain authority over Muslim lands, and Oza (2007) explains how colonial narratives have served to fulfil global supremacy agendas and to justify military action. For many Muslims living in the West (and not only them), such military disasters and humanitarian horrors evoke a strong desire, through political ideas, to express solidarity with oppressed Muslims. In the end, the globalized negative representation of Islam and Muslims aligns with other global geopolitical concerns involving borders, refugees, immigrants, security, the war in Syria, and so on (Nagel, 2016; Samari, 2016; Hopkins, 2007).

Consequently, in the West, anti-Muslim politics also impact on anti-immigrant and anti-refugee politics (Nagel, 2016). These anti-Muslim politics mainly frame public debates by speaking of 'Muslims in the West' instead of 'Western Muslims' (Dunn et al., 2007; Dunn & Kamp, 2009). Yet Muslims have lived in Western countries for decades now and should be accepted as full Western citizens. How many generations will it take before they are finally accepted as Western Muslims? The first generations were blamed for not speaking the language of their host country properly and for not participating in its social and political life. Today, in many Western countries, Muslims are from the second, third or even fourth generation of immigration and, even if they speak the language perfectly and engage in society and politics, are still denied belonging to their country.

Integration is the mutual process of being integrated and being accepted as integrated (Sardar, 2009). Questions about whether Muslims are capable of becoming integrated in fact reflect another reality. Indeed, I do not think that the debates are solely about integration, but are rather about assimilation. As many studies suggest, Muslims are already integrated into Western societies, whether culturally (Mills, 2009; Lewis, 2009), economically (Pollard et al., 2009) or politically (Glynn, 2009); however, the debates continue to fuel the 'Muslim problem' by questioning Muslim minorities' degree of integration, claiming that they are simply not making enough effort. To be considered as successfully integrated, Muslims are basically asked to erase or conceal a big part of their religious identity in the public space. Their degree of religiosity should not be used as a measure of integration; I think that better measures are an absence of contempt for them, equal treatment when searching for decent accommodation or jobs, an absence of discrimination and an explanation of minorities' presence. The majority sees this minority of Muslims as a problem principally because of their own fears and insecurities. Muslims should not have to hide their difference to be accepted by the majority as integrated, and can be proud of their presence in the West especially when it represents a major aspect of the national history. Finally, it is important to reassure the majority by helping them to have real faith in their own identity and, at the same time, it is vital to remove the minorities' complexes by showing that they do not have to be afraid of who they are, because, in fact, both can participate in transforming and creating their joint society.

National Islamophobia, secularism and Muslim identity politics

Islamophobia has qualities specific to the nation under study; nevertheless, it functions in the same way across the West, essentially by Islamizing any social, economic and political problems. Spatialized Islamophobia is nationalized, as some Western countries use nationalism and belonging to exclude Muslims from their national space. Indeed, as explained above, national Islamophobia is often related to the threatening transnational connection that immigrants and their descendants in the West maintain with their country of origin, as well as the perception of Muslims as Others who are disloyal to the West (Poynting & Perry, 2007; Selod, 2015). The relationships between the Muslim world and the West reveal that Muslim identity is connected globally, and important discussions (mostly in academic literature) focus on how Muslims experience inclusion and exclusion in shifting national, international and transnational contexts (Hopkins, 2004).

Another crucial argument, exploring the principle of secularism, is presented by certain Western countries. Kong (2010) provides a clear analysis of how secular states treat the Islamic faith in the modern world, explaining that religious considerations generally mean dogmatism and backwardness, while secularism means modernism associated with rational and scientific values. Secularism considers that religion should be excluded from civil affairs and from social and political matters; however, in countries where several religions coexist, the predominant or the politically dominant religion is decisive to the orientation of the entire society. In such societies, there is a double movement towards secularization in the dominant population and renewed religiosity in minority groups (Gale & Hopkins, 2009; Simon & Tiberj, 2013). This is why secularism is gaining ground in Western countries at the same time as there is increased visibility of religious minority groups. Moreover, public opinion is often mobilized to ensure the security of the nation-state as a secular, territorially defined, cultural and political form, warning against the perceived negative effects of religion on the secular state (McAuliffe, 2007).

In these contemporary debates, the threat of religion is essentially equated with the threat of Islam, regardless of country. Here, secularism exposes an important division between public and private spaces, and Muslim identities are uneasily incorporated into Western conceptualizations of the public sphere. Western Muslims are exposed to scrutiny in the context of the Muslim problem, because of the power of the politics and the media. These can influence how people see themselves and others through the particular representations that they convey. They play a significant role in the creation of social identities and have the power to define who and what a Muslim is. If people express fears about Muslims, it is because they are made to feel this fear.

Given the negative constructed image of 'the Muslim', Muslims have no choice than to react, either by compromising and negotiating their own identities in order not to draw too much attention to themselves or, conversely, by reaffirming their religious identity. On the one hand, some Muslims feel compelled to try to mask their Muslim identity as a result of the increase of Islamophobia, such as shaving off their beard (Hopkins, 2007) or acting and dressing differently in various contexts (Siraj, 2011; Dwyer, 2008; Carr, 2016; Najib & Hopkins, 2019). They try to reduce the extent to which they look like Muslims, because this negative identity created by the 'Muslim' label has contributed to many discriminatory acts. Muslim men and women worry about the social scrutiny to which they are subjected and the discrimination that they endure, which affects their everyday public and private lives.

On the other hand, the reinforcement of an oppressed identity can give rise to a political identity in response to injustice (Glynn, 2009; Najib & Hopkins, 2019). In the face of strong stigma, Muslims reaffirm their identity that is the most heavily oppressed, which is their religious identity. It is precisely about reclaiming their oppressed identity and giving it a status to reverse the stigma (Najib & Hopkins, 2019). Consequently, Muslims become proud of their religion and forge their own identity as Muslims, even if they also have other important facets. Phillips & Iqbal (2009) explain that Muslims have increasingly politicized their religious identities and that there are many ways to be a political Muslim. Some identify as national Muslims; others prefer to keep their religion private. Some directly associate with Muslim organizations; others prefer to work with non-Muslim associations (mostly with left party associations). Some are pious and political, while others are not religious and present themselves as 'vocal Muslims' (Modood, 2009).

Indeed, identity is a fluid multifaceted construct. A person has only one identity, but it is made up of several components combined in a unique mixture. That is why, in the main, Muslims living in the West have created a new hybrid identity (Dwyer, 1999, 2000; Sirin & Imamoglu, 2009). Gender, race, nationality, religion and other characteristics all contribute to the development of this sense of identity, and religion constitutes only a single facet (Datta, 2009; Nagel & Staeheli, 2009). Yet many Western Muslims create a hybrid positive identity that primarily asserts their religious heritage as Muslims and at the same time their civic and national pride as Western citizens (Sirin & Imamoglu, 2009).

Thus, while all these individuals are at times grouped under the umbrella term of Western Muslims, their degree of identification with Islamic and Western values varies. We cannot argue that, because a variety of people believe and practise the same universal religion, its practice is the same across space. Indeed, the form of practising a religion not only depends on geography and locality, but also on historic, cultural, political and socioeconomic realities. Muslims continue to be denied the opportunity to be represented or to represent themselves in terms of a complex, holistic and heterogenous humanity (Archer, 2009).

The stereotypes of a Muslim person assuming an identity that is primarily religious and incompatible with the secularized world are contradicted by the way in which, on a daily basis, Muslims juggle the relationships between religion and politics and between cultural difference and national belonging. The secularization of the modern Western world can lead to a loss of religious meaning (especially when nothing else is offered in its place) and to a focus on minority religious groups. Modood (2009) stresses that the group in Western societies most politically opposed to Muslims, or at least to Muslim identity politics, is not Christians, nor even right-wing nationalists, but the secular intelligentsia. The most fervent secularists want to impose their vision of the eradication of religion. They have somehow created a dogmatic vision that endangers not only Islam but all religions. Sometimes, even religious Christians become secondary collateral victims of Islamophobia.

Finally, secular policies have negatively impacted upon religious groups, above all Muslim populations who are more likely to suffer as a consequence of societal factors related to immigration, colonial past, terrorism, poverty and women's oppression (Abu-Lughod, 2002; Staeheli & Nagel, 2008; Allen, 2010; Dunn et al., 2007). Elshayyal (2018) clearly identifies an 'equality gap' experienced by Muslims who are not treated as equal to fellow non-Muslim citizens. The ways in which religious discriminations express themselves depict a specific Spatialized Islamophobia that interacts with global events and reflects a complex intersection and entanglement with other forms of discriminations related to gender, race, age and class. We need indeed to consider intersectionality (Crenshaw, 1991) to see how the gendered, racialized, aged and classed dimension of Spatialized Islamophobia applies, to varying extents, to Western nations and contradicts the liberal vision of the free Western world.

Challenging liberal Western values by combating Islamophobia

The increased visibility of Muslims in the West challenges the liberal system of contemporary Western societies. This liberal system promises liberty and equal opportunities for all, and this should provide good living conditions for Western Muslims. By contrast, they suffer significant stigma and injustice, sometimes even supported by State authority. In human geography, Gregory et al. (2011) define liberalism as 'the view that individual freedom should be the basis of human life' (p. 416). At the same time – along with others such as Castree et al. (2013) for example – they observe that this notion has been remarkably under-examined in relation to its rich history and ideas, as well as in comparison to the notion of neoliberalism, which is the subject of extensive critical analyses in Geography (Brenner & Theodore, 2002; Harvey, 2007; Ferguson, 2009). To be more exact, liberalism is a political philosophy that holds that the freedom of the individual to be a central means of governing our democratic societies and relates to many important principles, such as liberty, individualism, gender equality, civil rights, rule of law, free market, and so on.

Yet liberalism was born of the Enlightenment (in the 18th century), during which fundamental racist theories were developed. Liberalism can therefore also be used to legitimize reactionary racist ideas, whether they are ideas of the past or of today (Mondon & Winter, 2020). This can easily be observed in the trivialization of Islamophobic ideas in mainstream discourses and policies of our liberal democracies. Here, liberalism presents major contradictions to the values that it is supposed to defend (Losurdo, 2011). In the case of Islamophobia, which has become the mainstream form of racism in Western societies, the main contradiction is that

Muslims are the subject of targeted government objectives, often linked to their 'total' assimilation and very often enforced oppressively, in theory for the good of all. This practice, as well as this theoretical set of ideas, is indeed articulated through the activities of a ruling class that seeks to present its own interest as the universal interest of society (Harvey, 1976) by using liberal tropes as effective weapons against Muslims and Islam. Consequently, there is a clear 'Islamophobia of the liberal intelligentsia' (Akhtar, 1989; Weller, 2006), because otherwise, Islamophobia would not have the presence that it has in the media and politics without the advocacy of this dominant class which tries to obtain popular legitimacy supporting its opinion on Islam and Muslims.

At the end of the day, these liberal enlightened principles (of liberty and equal opportunities) are not miracle antidotes to curb systems of domination. On the contrary, they may even strengthen them (Losurdo, 2011; Parekh, 1992) because they are not applied equally to all, as evidenced by the many violations of the rights of Muslim populations. Likewise, liberal principles of open competition and freedom of expression create deep divisions among other populations, even though these principles have become widely accepted social norms. Indeed, the fact that the extremely rich become richer and the very poor are likely to become poorer (Oxfam, 2018; Piketty, 2013) is no longer even shocking, as is the fact that White elites (mainly men) use racist arguments to avoid losing their privileges (Mondon & Winter, 2020). Ultimately, liberalism can lead to a higher level of socioeconomic inequality and xenophobia (Morgan & Poynting, 2012; Weller, 2006; Mondon & Winter, 2020) that may threaten both Muslims and non-Muslims.

Regarding xenophobia, systems of domination have usually affected vulnerable minority populations in Western history, such as Black, Jewish, Irish[2], immigrant and refugee groups. The West has claimed intellectual superiority over other civilizations through its politics of domination (from slavery to colonization and Westernization) (Grosfoguel & Mielants, 2012). For several decades now, the emphasis has been focused on the figure of the 'Muslim' (Weller, 2006), and many questions have been raised on how to accommodate Islam in Western democratic societies, while current leaders (whatever their parties) share global stigmatized discourses on Islam and Muslims in the name of liberal values. Indeed, Islamophobia is one of the prime factors to be associated with the success of far-right and populist movements (Abbas, 2019; Renton, 2003), alongside other concerns related to immigration, welfare chauvinism, multiculturalism, the European Union, the globalization, and so on (Mudde, 2007). They have successfully shifted their focus from anti-Semitism, through racism against Black people, to this new central racism against Muslims.

This success can be personalized around the former president of the United States, Donald Trump, who argued for example that 'Islam hates us' (Hassan, 2017). Such negative narratives are structured within a global political discourse that can definitely lead to Islamophobic attacks against Muslim populations and institutions, and the US policy actively participates in this. Discriminatory national security policies towards racialized Muslim communities have been implemented without really causing a major social crisis (Ali, 2012), and political actions inspired by the 'global war on terror' have been used to justify a new form of surveillance against Muslims, as well as ongoing discrimination (Fekete, 2009; Poynting & Mason, 2007).

The liberal system observed in Western countries has a disturbing tendency to associate itself with fascism, because White supremacy is currently empowered and has driven Western policy over the past few years (Cainkar, 2019; Mondon & Winter, 2020). For example, by exploiting security rhetoric and the inability of governments to raise the minimum wage, populist speakers advocate the protection of their own interests and keeping jobs for their own people, even if this means excluding racialized minorities from their own country. Since there is no strong opposition to these policies, populist reactions to the political status quo are increasingly visible in Western countries (Worth, 2014). This populist success is mainly explained by the growing contempt for traditional establishment institutions right across the Western world. We can observe it not only in the so-called strongest country in the world with the election in 2016 of a populist president, but also in Europe with, for example, the implementation of Brexit in the United Kingdom or the election of the French president Emmanuel Macron, presented as an 'outsider' and a 'reformer' during his presidential campaign (from neither the traditional right or left political parties).

This unequal shift that is ravaging our Western societies represents the real political problem, rather than the 'Muslim problem'. It is therefore more than necessary to refocus priorities onto questioning the allegedly egalitarian and democratic provision of Western countries. The triumph of unequal values in various Western countries is evident, and the liberal system needs to take responsibility for the desolation of Western democracy. Liberal democracies have themselves nurtured belief in the Muslim threat in order to divert attention from their inability or unwillingness to find proper solutions for social inequalities (Mondon & Winter, 2020). This liberal system has given rise to political means of domination, so that they can present authoritarian, even totalitarian, forms. Indeed, a totalitarian regime assumes that its world is the whole world. Consequently, the sensibilities of those from elsewhere, who have other priorities, do not matter, because this regime precisely requires them to be different to who they are (Todd, 2015).

Rosanvallon (2015) goes so far as to say that, even if our Western regimes are said to be democratic, we do not live in a democracy. Western countries offer obvious elements of freedom that we must use and exploit, but the word 'democracy' hides institutional forms and practices that present obstacles to people's emancipation, especially Muslims. Muslims are today challenging liberal democracies as well as their rule of law. Many violations have been recorded over the past decades. Popular democratic initiatives such as the prohibition of minarets, the ban of the Islamic veil in public schools and the surveillance of suspected Muslims and mosques have occupied the public sphere, sometimes using neo-Orientalist and imperialist arguments about Islam and Muslims. Liberal political discourse in Western countries often revolves around the following Islamophobic tropes: Muslims come from backward cultures and do not belong in Western countries; the Islamic veil oppresses passive Muslim women who need to be saved; and Muslim men are terrorists (Hopkins, 2009; Dwyer, 1999; Staeheli & Nagel, 2008). Much rhetoric following terrorist attacks, Donald Trump's presidency, the implementation of Brexit and the rise of far-right groups has internationally promoted a climate in which Islamophobic actions have become normal and even legitimized (Zempi & Awan, 2019).

This liberal programme comes with great injustice: it is indeed acceptable, in the name of freedom of expression, feminism and security, to stigmatize Muslims, to insult them, to control their dress and religious buildings, to racially profile them (even their children), to brutalize them and therefore violate their civil liberties and rights. Muslims are framed as 'evils' and 'enemies' in an elite-engineered moral panic (Bonn, 2012), naturally encouraging people in power to eliminate this threat (Cohen, 1972; Garland, 2008). These violations must end, and strong reactions must be raised before the worst periods in our history come back.

According to many scholars (Said, 1978; Bunzl, 2007; Bangstad & Bunzl, 2010; Klug, 2012; Bell, 2018; Hafez, 2016), this phase of Islamophobic stigmatization, discrimination and violence, to a certain extent, presents similarities with the phase preceding the Second World War and the genocide of Jews in Europe. Indeed, when the Jewish problem was constructed and worsened, the Jew was stigmatized. Suspicious intentions and the desire to infiltrate, control and conquer Western societies and in particular Europe were attributed to the Jew, just as the theory of the great replacement of European populations by Muslim populations does today (Danis, 2016). This phase of stigmatization was obviously faster and more violent for the Jews in the 1930s than it is today for Muslims. But many other factors of that time, such as the economic crisis, the rise of populism and the witch-hunt of critical scholars, are also observable today. They contributed to the anti-Semitic hysteria and led to the unspeakable atrocity, which was not so long ago. This parallel between anti-Semitism and Islamophobia, although very sensitive, remains an important subject for discussion if we are to prevent and combat future tragedies, because the real goal here is to recognize racist mechanisms and to counter any threat before a comparable situation arises (Schiffer & Wagner, 2011; Hafez, 2016). Besides, any increase in Islamophobia necessarily leads to more anti-Semitism (Mayer et al., 2019) and other forms of racism (Poynting & Noble, 2004). This is why today many voices globally are raised against Islamophobia to denounce in the same dynamic manner any form of injustice.

The fight against Islamophobia cannot be dissociated from the fight for more social justice in general. It is recognized that justice is indivisible: it is not possible to seek justice for a single group without seeking justice for all the people who suffer from the various injustices of our societies (Davis, 2016). Therefore, Islamophobia is also a feminist issue, an anti-racist issue, an anti-antisemitic issue, and so on. The idea here is to honour the values of equality and human rights that Western liberal countries are supposed to defend. Liberalism allows our Western governments to spread their contemporary 'truth' (Foucault, 1980), which can 'Otherize' Muslims (Kumar, 2012; Weller, 2006). But how can these governments protect the interests of their citizens by dividing their own peoples and countries? There is a strong need to change how we are conducted and directed. The absolutely positive idea of the West (Bonnett, 2004) needs to be criticized and challenged, as the West can be contradictory, exploitative, unequal and intolerant (Hopkins, 2009). Civil disobedience to this type of liberal thinking through social and popular protests may be a possible solution.

The best challenge to our Western democracies' liberal contradictions regarding Islamophobia (Losurdo, 2011; Isakjee et al., 2020; Weller, 2006; Mondon & Winter, 2020) seems to be from veiled Muslim women, since they represent the antithesis of the ideal Western citizen (Razack, 2008; Carr, 2016). Veiled Muslim women are seen as submissive victims whom the West needs to save from their backward culture (Razack, 2008; Spivak, 1988; Mohanty, 2003; Fernando, 2009; Scott, 2007). But if their freedom to wear whatever they want is not respected, veiled Muslim women do not need to be saved like this. Often, Muslim women's bodies and outfits play a central role in the so-called liberation and alleged emancipation of women and the construction of national identity. The body of the Muslim woman becomes the focus, as if it is a malleable and docile entity that may be examined and disciplined at will. In this liberal context, to be considered as a good citizen with good behaviour, the Muslim woman must conform to Western norms, notably by unveiling (Ahmed, 1992; Gökariksel & Mitchell, 2005; Carr, 2016). Thus, a veiled Muslim woman is stereotyped as part of 'failed' integration, while the Muslim woman who does not wear a *hijab* is assimilated and symbolic of a 'successful' integration (Fernando, 2009; Listerborn, 2015).

But Muslim feminism does exist, and it is becoming increasingly powerful (Bullock, 2005; Cheruvallil-Contractor, 2012; Scharff, 2011; Zimmerman, 2015). Veiled Muslim women control their own minds, they are their own agents and they can also be feminists. If these women can be supported to take up opportunities to promote change and resistance, it might definitely challenge liberal values (related to freedom, individualism and feminism) more akin to dominant global discourses, in part because of Western imperialism (Grosfoguel & Mielants, 2012). In fact, all Muslims, not only veiled Muslim women, should be treated as deserving of a legitimate and full place in the West. Their inclusion within society should be regarded as an important criterion of whether a society is indeed democratic, egalitarian, republican and multicultural (Modood, 2009).

Overall, all these injustices overlap and result from the liberal policies of those in power over the past decades. I have previously explained how the liberal system tends actually to promote greater economic injustice and authoritarian nationalism, yet I must reiterate here that I am neither an economist nor a political scientist, so I will not push this analysis further in this book. I am a geographer and I can thus observe that no space is neutral and that it can be used as a means to achieve liberal aspirations, such as spatial exclusion, that economically and politically divide citizens. Citizens are not only *Homo oeconomicus* and/or *Homo politicus* (Brennan, 2008; Nyborg, 2000): they are also *Homo geographicus* (Sack, 2007), physically located in a specific socio-spatial environment that they must share. Liberal policies create a liberal country that excludes its scapegoats from national belonging to their own country and spatially marginalizes them (Mehta, 1999).

The concepts of nation, state, country, nationalism and belonging are widely used in the discipline of Geography, which seeks to understand spatial organizations and implications as well as the complex relationship between various spaces and their residents. Islamophobia, like fear (Pain, 2009), is analysed through powerful metanarratives that circulate on a global scale and clearly question the space and the place occupied by Muslims; which may become a leading research analysis in human geography. From global to national levels, Spatialized Islamophobia travels across borders and adapts itself to the various contextual microcosms while spreading negative ideas about Muslims and their supposed will to take control of the religious and political space in the West. People's fear that the West will be Islamized seems irrational, yet this is at the forefront of public debates and dominates public space. These ideas are exclusively to divide up space (and therefore people) between Muslim and non-Muslim spaces, particularly in terms of public/private spaces, secular/religious spaces, inclusive/exclusive spaces, White/coloured spaces, and so on.

These spatial divisions have been recognized by geographers as problematic (Staeheli, 1996; Bondi, 1998). The spatial reading of Islamophobia precisely implies a binary spatial opposition 'here versus there' (Clayton, 2009; Hancock, 2015) that mirrors the political opposition 'us versus them' (Saeed, 2007; Little, 2016), defining 'us' as superior, White, non-Muslim bodies, modern and secular citizens, and 'them' as inferior, Black/Brown, Muslim bodies who are foreigners in Europe, North America or Australia and should remain in the Middle East, South Asia or Africa (Najib & Teeple Hopkins, 2020). This liberal 'truth' – that political leaders of these past decades have summarized as 'the good versus the evil' (Fish, 2001; Mamdani, 2004) – aims to dictate the gaze we must have towards the Other, and thus feeds anti-Muslim racism (Kumar, 2012).

The first function of racism is indeed to divide people, and the spatial reading of global Islamophobia divides spaces by referring 'there' to Muslim countries and in particular to war zones in the Middle East for example, while national Islamophobia refers 'there' to disloyal spaces occupied by the 'othered' Muslim (i.e. the enemy within) who wants to impose his/her religion spatially and who directly brings conflict into the country. This spatial reading is even clearer when analysing and comparing in greater depth specific Western nations (and when zooming in to a finer scale, as in the next chapters). It is important to challenge this spatial vision of Islamophobia by showing that, in fact, Muslimness is not only associated with the 'there' but it is also a part of the 'here' (Carr, 2016). Indeed, the religious heritage of Western Muslims may lie in their non-Western background, but just because Islam is a minority religion in the West does not mean that it is a religion foreign to such countries or that it cannot be a Western religion, alongside many others.

Finally, analysing Islamophobia in Western nations is deeply important because, if Western values are capable of yielding the fundamental values of freedom, democracy, modernity, individualism, secularism and gender equality (Grosfoguel

& Mielants, 2012), why can't Muslim populations find their full place in such countries. Why don't liberal rights apply to them, and why are their claims for more equity unwelcome? Islamophobia threatens Muslims' everyday lives in various Western countries, and issues of religious intolerance, racial discrimination and ethnic identity are just as important in France as in the United Kingdom or Germany, Denmark, the United States, Canada and Australia. Each country has a different conception of these issues and how Islamophobia functions depending upon its history, political culture and legal system (Modood, 2009; Zempi & Awan, 2019). It is particularly interesting to focus on the French case that ignores racial, ethnic and religious identities in connection with its principles of *laïcité*, as I do in the next section, as well as on the British example where religious equality is less valued than gender and racial equality (Modood, 2009; Allen, 2010).

Particular instances of French Islamophobia and British Islamophobia

At the global scale, European anxieties and panic about Muslims are significant (Morgan & Poynting, 2012), and this book presents the phenomenon of Spatialized Islamophobia while making practical observations of the French and British contexts. France and the United Kingdom are both important countries with a large number of Muslim populations (Duncan, 2016; Hackett, 2017) and a high rate of Islamophobic acts (Ray et al., 2014); however, they demonstrate two different Islamophobias (Najib & Hopkins, 2020). The French form differs from the British with regard to the type of anti-Muslim acts, where they occur and who they affect. Collecting quantitative data from the CCIF association in the French context and from the Metropolitan Police, as well as the British associations MEND and Tell MAMA (Measuring Anti-Muslim Attacks) helped me to draw up an overview of the functioning of Islamophobia in both countries.

On the basis of an in-depth quantitative analysis of data from 2015[3] in France and the United Kingdom (Najib & Hopkins, 2020), I briefly explain that in France Islamophobic incidents are mainly discrimination in public institutions, whereas in the United Kingdom they are mostly verbal abuse on public transport (ibid.). In both countries, Islamophobia mainly affects veiled Muslim women who are most likely to be students in France and shoppers or passers-by in the United Kingdom. They are overwhelmingly from a foreign ethnic background (i.e., North African in France, and South Asian in the United Kingdom), and are more likely to be from a disadvantaged socio-occupational category (ibid.), in accordance with the general situation of the Muslim populations in both countries (Gale & Hopkins, 2009; Pollard et al., 2009; Simon & Tiberj, 2013). Consequently, in both the typical profile of victims takes on the known pattern of domination in contemporary societies regarding sexism, racism, ageism and classism.

Overwhelmingly, the nature of the Islamophobic acts and where they occur are linked to specific French and British policies. First, France has a problem with religion in general and with Islam in particular. French history reveals a clear desire to break with religion and religious guilt, and as a result, France has become highly anti-religious. It has been at the forefront of implementing laws against religious dress in European nations in the name of secularism. A secular state normally seeks to be rationally independent of any religion while promoting the free exercise of all religions. But the French *laïcité* – initially developed in the early 20th century to separate the Catholic Church from the State and, indeed, to protect minority religious groups (Baubérot, 2012) – was revised in the early 2000, after several *affaires du foulard* (headscarf incidents) in local schools to implement a law (15 March 2004) to prohibit wearing ostensibly religious items in public institutions. This law was motivated by the *hijab* (Delphy, 2006) and affects mostly veiled Muslim women, as they must remove their headscarf if they are students in primary and secondary schools or if they perform a job representing the State (in a town hall, for example) (Glynn, 2009; Najib, 2020a). Without considering the emotional violence that this law can cause, French officials, in formulating the ban, asserted the primacy of *laïcité* over people's right to express religious difference (Bowen, 2007). From this local start, multiple anti-Muslim policies and discussions have now affected the entire national space. The 2004 law was the starting point and supporting principle for further policies and discussions stigmatizing Muslims in France (e.g. the prohibition of the full-face veil, police surveillance, discussions on a *hijab* ban for nannies and mothers participating in school trips, and also in universities or on beaches with the *burkini* affair, etc.).

France appears to be the test bed for Islamophobia in the name of secularism, developing influential Islamophobic tools that inspire other European countries (such as Denmark, the Netherlands, Switzerland, etc.). Therefore, the principle of *laïcité* has now deviated from its original meaning and been transformed into a weapon against Muslims. As a result, French Muslims feel betrayed by the failure of the French authorities to guarantee the 'right to difference' because of how secularism is now defined in France. That is to say, religion no longer has a place in public space (Esteves, 2019). This 'political enterprise' in France (Lorcerie, 2005; Teeple Hopkins, 2015) has ultimately led to the hyper-normalization of Islamophobia and a multiplication of Islamophobic incidents, generally manifested in institutionalized discrimination mainly due to the false interpretation made in the 2004 law (Najib, 2020a).

Compared to France, Islamophobia in the United Kingdom is very little expressed through institutions and law. British pragmatism contrasts with the French policies, which are built on the secular system (Glynn, 2009). Indeed, governmental actions vary by nation, and in multicultural Britain, any feuds over headscarves are relatively rare because the government protects the plurality of ethnic-religious identities. But the place of Islam in the United Kingdom has been much discussed, and questions about Muslims and their belonging, self-segregation and faith schools shape contemporary public debate on religious difference (Dwyer & Uberoi, 2009; Modood, 2007), necessarily leading to an increasing number of verbal Islamophobic attacks in public (Najib & Hopkins, 2020). Nagel & Staeheli (2009) clearly explain that Muslim identity has become public and political, while for British Muslims religion is a matter of personal faith.

These observations lead us to question the various ways in which Islamophobia functions in each country. In France, we can undoubtedly read the impact of the

2004 French law as a top-down process that emanated from the State, and not necessarily from the people. Conversely, in the United Kingdom, this phenomenon is rather a bottom-up process, primarily from individuals under the influence of negative media representations (Ameli, 2004; Saeed, 2007). Here, the particular perception of secularism in France leads us to note a specific type of Islamophobia in a country obsessed with the Islamic veil, normalized through its ban in public institutions, in contrast to the British context where there is no such principle of *laïcité* governing the functions of schools and institutions and questioning the wearing of headscarves. We can nevertheless observe in the United Kingdom that the political sphere has become increasingly important and has greatly contributed to redefining the ways in which Muslim identity should be embodied and expressed in public spaces (Modood, 1997; Parekh, 2000).

Ultimately, the functioning of French Islamophobia and British Islamophobia follows the State model: the French Republican model bases its principles on equality among citizens and their social mix, while the British multicultural model recognizes all its inhabitants' differences and does not necessarily encourage the intermingling of contrasting communities. Consequently, the zones of contact in Britain may be subject to more violence when such communities meet than in France (to be explored in more detail in following chapters). Finally, this work challenges not only French policy (fostering institutionalized discrimination) but also the way British society behaves (especially in zones of contact). These two State models have a direct impact on the spatial dynamics of Islamophobia observed in France and in the United Kingdom, as well as on its gendered and racialized dimensions.

The gendered dimension of Islamophobia is important both in France and in the United Kingdom (Najib & Hopkins, 2020). Secular laws affect Muslim women who wear a headscarf to a greater extent than Sikh or Jewish men who also may wear a turban or *kippa*, and Islamophobic attacks and interpersonal assaults in public spaces tend to target visibly Muslim women (Hancock, 2015; Najib & Teeple Hopkins, 2020; Gökariksel & Secor, 2015; Listerborn, 2015). To be precise, in the United Kingdom the typical aggressor is a British White man between 25 to 59 years old (Copsey et al., 2013; Tell MAMA, 2016), whereas in France the aggressor is as likely to be a woman as a man (CCIF, 2011; Najib & Hopkins, 2020). Therefore, in the United Kingdom, the gendered dimension of Spatialized Islamophobia seems a matter of male domination and control in public spaces, while in France Islamophobic acts can also be by certain French feminists who are *de facto* opposed to the *hijab*. These women tend to resist the veil without considering the voices of the wearers, the great majority of whom explain that it is part of their free identity.

These feminists usually agree with the *hijab* ban, arguing that a veiled Muslim woman cannot be a feminist and a free woman (Scott, 2007; Fernando, 2009). A double standard emerges, showing one group of women attacking another in the name of women's emancipation. These feminists usually highlight the conditions of other women in authoritarian countries like Iran, for example, who are forced to wear the *hijab*, but how can they meaningfully differentiate that obligation to veil from the obligation to unveil? Both Iran and France's laws on religious dress treat women bodies as sites of State operation and legitimacy (Dabashi, 2012), yet France is known as a democratic country whose national motto is *Liberté, Égalité, Fraternité* (liberty, equality, fraternity). The country is represented by the allegorical female figure of Marianne (like England's Britannia), who is a symbol of freedom and democracy, and she is often made symbolically opposed to veiled Muslim women (Mukherjee, 2015), better to exclude them from Frenchness and from feminism.

Any type of feminism that does not support freedom in terms of either dress or religious liberty is likely to become trapped in versions of racism and class ideologies. Indeed, universal feminist struggle consists of transforming society not only in terms of sexism, but also in terms of racism, classism, and so on, of course including Islamophobia. Universal feminism fully supports social justice and thus the freedom for women to wear as much or as little as they want (Feltz, 2006). Therefore, this type of feminism, which excludes Muslim women from universal feminism, is known as 'White feminism' (Frankenberg, 1993). This removes any agency from veiled Muslim women (also from racially minoritized women) for their supposed own well-being, and seeks to unveil them. This obsession with unveiling has a long history among Western colonialists and imperialists, who want women to conform to the dominant norm (Macdonald, 2006; MacMaster, 2012). White feminism has an important colonial dimension and dictates how minority women should dress and behave. These narratives have drawn criticism, especially from Black feminists, who argue that White norms are being used to judge other cultures and then feed racism (Carby, 1982; Spivak, 1988; hooks, 2000).

Islamophobia thus also has a racialized dimension. This study's findings reveal in each country a strong focus on a highly stigmatized ethnic minority (Najib & Hopkins, 2020), and refer to their current relations with their former colonial empires (Hancock, 2009). Interestingly, cultural diversity is discussed differently in both countries, mainly due to the lack of racial and religious data in France and its imperial tendency to incorporate its former colonies politically and culturally into a greater France (Glynn, 2009). Indeed, French colonization proclaimed itself 'assimilatory', while British colonization considered itself as 'respectful of cultures' (Balibar & Wallerstein, 1991). The French position would be that justice is blind to difference, a precondition for making universal statements (Hancock, 2009). There is therefore an important reluctance to discuss race, ethnicity and religion in France (Amiraux & Simon, 2006), but ignoring racial, ethnic and religious identities does not mean that the related problems of exclusion, racism and fragmentation do not exist or are less numerous. They are just more hidden and therefore more difficult to combat.

In the United Kingdom, the situation is clearly different: social justice is unthinkable if it does not take into account differences. Ethnicized and racialized minorities are increasingly regarded as having religio-cultural characteristics (Abbas, 2009). For example, due to the growing prominence of Islamophobia, the South Asian community in the United Kingdom (mainly composed of Indians, Pakistanis and Bangladeshis) has largely been perceived as a Muslim population

(Allen, 2010). For Muslims, this shift has been legally problematic, since UK racial protection affords protection to only the mono-ethnic religious groups of Jews and Sikhs (Modood, 2009; Allen, 2010).

To be precise, until 2006, legislation failed to protect Muslims either on the basis of their religion or as a racial group, because they could not be recognized as such, unlike Jews and Sikhs for whom there is an established legal precedent for inclusion within the meaning of 'racially' abuse, as per the Race Relations Act 1976. To counter this important difference, the Racial and Religious Hatred Act 2006 was introduced, followed by the Equality Act 2010.[4] These protect religious identity, among many other identities (such as gender, race, disability, etc.). But in practice, Jews and Sikhs continue to have recourse to racial discrimination provisions to defend themselves against racial abuse, unlike Muslims (Meer, 2008). In addition, the government remains sluggish and equivocal about adopting a formal definition of Islamophobia and recognizing it as anti-Muslim racism. It remains therefore difficult for Muslims to pursue any Islamophobic perpetrator on the basis of racial abuse under existing incitement law. Finally, while discrimination against *kippa*-wearing Jews and turban-wearing Sikhs is considered to be illegal racial discrimination, Muslims have no choice but to prove that their experience of discrimination is either directly against their ethnicity (Asian, Black, etc.) or their religious appearance (*hijab*, beard, etc.).

In France, too, there is a shift from race to religion, and those in the 'Arab' community (more exactly North African or Maghrebi) are now more likely to be perceived as Muslim, causing a shift in the discrimination and abuse, from Arabo-phobia to Islamophobia. There is nothing in French law to protect racial and religious minority groups since they are not even recognized. France is not a multicultural country; it is a Republican country in which ethnic and religious origins do not exist. In practice, it is obviously impossible to ignore racial and religious groups in everyday spaces. This blindness concerns only official census data and law protection, and Muslims are much targeted and attacked in the public sphere.

To take the example of the right to blaspheme, this seems a French priority especially after the Charlie Hebdo attacks in 2015. Todd (2015) denounces this relentless affront on Islam which is the religion of one of the most vulnerable populations in France. Making blasphemy a national project is, according to him, a scam that primarily seeks to target this minority religion and its followers rather than to seek a respectful compromise. This national project of blasphemy was even reinforced in October 2020 (just after the terrorist murder of a schoolteacher outside Paris and three people in a church in Nice) by means of a bill on anti-separatism[5] (also known as 'Islamist separatism') that President Macron chose to implement to defend *laïcité*.

Roy (2020) clearly explains that such measures by the French government have not prevented terrorist attacks, not since the bombing of the Paris Metro in 1995. This bill, which became law in the summer of 2021, is itself at odds with French Republican values and therefore presents significant liberal contradictions (Losurdo, 2011). First, it seeks to defend the right to blaspheme at the expense of the other liberal values that it is supposed to defend, such as freedom of religion, belief and opinion (Roy, 2020). Second, separatism, more generally, has always been the structure of French policies towards postcolonial subjects who, for example, suffer from significant economic, educational and territorial separatism. The fight against separatism should promote a better inclusion of Muslims in French society at all levels, instead of making them understand that to be better included they must accept being insulted (through blasphemy) in the name of freedom of expression. This is neither a respectful way to promote freedom of speech (far from being total in France) nor to engage in criticism of Islam.

By contrast, in the United Kingdom, Modood (2009) explains that there is a greater understanding of blaspheming without giving gratuitous offence to Muslims, probably thanks to John Lockes' classic work (1689, as cited in Weller, 2006) on tolerance in a liberal democracy. Criticism of Islam is possible, but a clear distinction must be made between unfounded criticism (defined as negative, essentializing, insulting or disrespectful) and founded criticism (respectful of individuals, and open to dialogue and serious enquiry into the diversity of Islam).

All told, these examples from France and the United Kingdom show the urgency of considering the racial issue and racism in the study of Islamophobia. Much has been written in academia on the racialization of Islamophobia (Halliday, 2003; Naber, 2008; Allen, 2010; Sayyid & Vakil, 2010; Scott, 2007; Dunn et al., 2007; Hopkins, 2004), and the political sphere must follow. Indeed, linking Islamophobia with racism allows us to better understand that Islamophobia is as unacceptable as any other kind of racism, contrasting with the argument that Islamophobia is only a simple criticism of Islam. Islamophobia is more than that; it targets the Muslim identity.

In 2018, an attempt to educate the British government on the racialized dimension of Islamophobia was led by the APPG (All-Party Parliamentary Group) on British Muslims. Following various consultations with experts such as academics (including the findings of the SAMA project), policy-makers and community activists, this group recommended that Islamophobia should be regarded as a type of racism that targets expressions of Muslimness or perceived Muslimness. Despite being widely accepted by the academic world (Zempi & Awan, 2019), this definition was rejected by Theresa May's government under pressure from several groups, notably the anti-terror police (ITV Report, 2019). Here, a new opportunity to address Islamophobia effectively in the United Kingdom seems to have been lost (Allen, 2019). In France, 'Muslim' is an unrecognized category since, as explained above, the government refuses to call into question its constitutional principle that the French people are one and indivisible. For this reason, it is challenging to make claims in France on behalf of Muslim populations, and difficult for them to organize themselves.

The UK and French governments should recognize the racialized dimension of their country's Islamophobia and themselves monitor quantitatively all anti-Muslim acts, if they are to eliminate Islamophobia. Carr (2016) explains that anti-Muslim data are collected by NGOs, which to some extent play the role of the

State. Why do the French and British states leave the fight against anti-Muslim racism mainly to NGOs? Is it because Muslims do not conform to their hege-monic norms (Carr, 2016; Lentin & Titley, 2012), or because the governments think Islamophobia does not exist or is not a priority, or even because Islamo-phobia has become acceptable racism? If governments themselves do not measure anti-Muslim racism efficiently, then its existence becomes difficult to prove, and therefore this form of racism becomes harder to challenge and eradicate. The French government even crossed a red line with the dissolution in November 2020 of the CCIF, the main NGO documenting French Islamo-phobia and providing support to victims of Islamophobia. The bill on anti-separatism was partly behind this dissolution, showing that France does not allow the defence of its Muslims. It was not a decision that was taken on evidence of any illegal action incriminating the CCIF, but more an ideological act seriously calling into question the fundamentals of the rule of law and democracy in France. Yet, Muslims (or those who are supposed to be) con-tinue to need to be defended since they are affected by their visible Muslimness.

Indeed, it is the physical identity that is clearly targeted (Hopkins, 2004), and perpetrators sometimes mistake a range of minority ethnic and religious people as Muslims, notably Sikhs (Hopkins et al., 2017). Therefore, Islamophobia targets those who present visible markers of Muslimness, as several studies from around the world have emphasized (Allen & Nielsen, 2002; OSF, 2011; Zempi & Chak-raborti, 2014; Carr, 2016; Najib & Hopkins, 2019; Garner & Selod, 2015; Bayrakli & Hafez, 2019). Muslims can be targeted because of their clothes, headscarves, beards, names, and so on, in the same way as phenotype and skin colour in traditional White supremacist racism (Cainkar, 2019). Obviously, a Muslim can also be attacked on the basis of his/her race, but Islamophobia also functions in a form that targets Muslim appearance. Even White converts are attacked, because of their Muslimness, and they become racialized and 'culturally others' (Van Nieuwkerk, 2004, Moosavi, 2015). For example, when a White woman of French or British origin converts to Islam and decides to wear the *hijab* (a strong marker of Muslimness), she is assumed to be from a foreign background. Attackers may tell her, 'Go back to your country', when she may be a native with a long French or British heritage.

When people's identity is not accepted or is the subject of multiple discrimina-tions, the feeling of exclusion can be deeply violent for those who consider themselves as belonging to this special category. Ahmed (2000) explains that an individual becomes a stranger when we begin to recognize him or her as such and when he or she begins to feel that way. Thus, it is common for some converts to feel foreign and to distance themselves from the majority population, especially after deciding to wear the veil. Muslims experience strong hostilities because of their physical appearance, their dress and the racialized constructions of their faith. Given its similarities to the classic manifestations of racism, Islamophobia should indeed be classified as such (Carr, 2016), not only in these specific cases of France and the United Kingdom, but also elsewhere.

Conclusion

Islamophobia is one of the most worrying planetary prejudices of our contemporary era. It is a global phenomenon, and the Western world founded on liberal principles should react first, as anti-Muslim racism is not only a serious threat to Muslims but also to Western democracies. Indeed, Islamophobia does not concern only Muslims; it is a human rights issue that concerns everyone. Everyone should stand against this anti-Muslim racism and stop acting as if this issue should not be discussed, because nowadays it is basically irresponsible not to talk about Islamophobia, whether in India, China, Myanmar, the United States, France, the United Kingdom, Austria, Denmark, Australia, Turkey, Tunisia... Too many people in the post-9/11 era have suffered significant abuses and violence or have even lost their lives in their own locality because of global Islamophobia (Zempi & Awan, 2019), whether it involves Muslim-minority contexts or not, and Western countries or not.

Through the examples of France and the United Kingdom, it is possible to better understand the national mechanisms of Spatialized Islamophobia in these countries. For instance, French Islamophobia seems to be an institutionalized consequence of the 2004 law (Najib & Hopkins, 2020; Najib, 2020a) which, because it is poorly applied, affects users of public services. Therefore, concreate measures against an abusive application of the 2004 law can be triggered (training, public education, etc.). In the United Kingdom, the Islamophobic acts mainly take the form of verbal abuse of Muslims on public transport. Therefore, local transport companies should introduce surveillance (police patrols, CCTV, etc.) to counter such hate crime.

National Islamophobia means that being a Muslim in France or the United Kingdom can be a daily challenge and that Muslims' place is secondary. Such conditions are not a random circumstance; they are rooted in liberalism. This liberal anti-Muslim racism is becoming more and more accepted, and as long as our Western governments are unprepared to break with their liberal regime, all strategies to reduce Islamophobia (and more broadly inequality) will necessarily reproduce unevenness. Put simply, both Muslims and non-Muslims need to criticize and challenge the liberal contradictions of nations that claim to limit, contain and reduce the inequality that the governments have paradoxically engendered through economic deregulation and racial blindness (Cassiers & Kesteloot, 2012; Wacquant, 2015; Losurdo, 2011; Isakjee et al., 2020; Mondon& Winter, 2020, Najib, 2020b). On the one hand, in order to regain their agency and develop a voice independent of any outside power, Muslims must empower themselves (their identity, institutions, leadership, etc.). And on the other hand, non-Muslims must keep the authentic issues that affect them on a daily basis at the centre of their political and media debate and not allow the anti-Muslim discourse to detract from pressing socio-economic problems (such as increased unemployment, poverty, health, insecurity, etc.).

To end this chapter on an optimistic note it is important to remember that, even if anti-Muslim acts continue to increase after each terrorist attack, a feeling of solidarity between communities (in particular between Muslims and non-Muslims) may also grow (especially in France) (Mayer et al., 2019). Thus, it is possible that society is less divided than the incidence of global and national Islamophobia might indicate. At a finer scale, people can distinguish what they see on television and in newspapers from what they actually experience in their everyday local spaces.

The priority is certainly to build bridges between communities and challenge the social divisions erected by policies and discourses (Phillips & Iqbal, 2009), as well as achieving better social cohesion among all the actors fighting against Islamophobia. These include those who provide a relevant definition; those who describe the phenomenon; those who directly support victims by providing legal assistance; those who attempt to have Islamophobic incidents recognized as separate hate crimes; those who measure and monitor the data; those who provide a more positive image of Muslims in the media; those who try to engage in political lobbying and electoral activism; those who raise awareness in the media and parliaments; those who concretely name and shame Islamophobic violations to international organizations; those who directly interact with Muslims in mosques, etc. If all these actors work together, the impact of such work will also be felt in more localized contexts, such as our cities and districts. Indeed, Spatialized Islamophobia also relates, to an extent, to urban and infra-urban spaces (Listerborn, 2015; Itaoui, 2020; Fritzsche & Nelson, 2020; Najib, 2020a). This is what the next chapter explains.

Notes

1 The tower that provides a visual focus and is used for the Muslim call to prayer.
2 It is fascinating to see how Jewish, Irish and Italians were racialized as non-White, in the past, and not today (Gilman, 2000; Sullivan, 2006; Ignatiev, 1995).
3 The data are from 2015 (the most recent available at the time of the investigation).
4 The Equality Act 2010 aims to protect people against discrimination, harassment or victimization in employment, and as users of private and public services based on nine protected characteristics: age, disability, gender reassignment, marriage and civil partnership, pregnancy and maternity, race, religion or belief, sex and sexual orientation.
5 This draft law plans to reduce displays of the Islamic faith in public life, to control religious and cultural associations, to cancel the right to choose a doctor of the opposite sex, to ban home schooling apart from for health reasons, and so on.

3 Urban and infra-urban Islamophobia

Introduction

Islamophobia is not only a global phenomenon: it is also locally targeted and can be observed in cities and districts (Najib & Hopkins, 2020; Itaoui, 2020; Najib, 2020a). Indeed, Islamophobia is practised in localized ways and highlights specific areas, places and streets. On the urban and infra-urban scale, the spatialization of Islamophobia needs both quantitative and qualitative materials, and connects to a varied literature on residential segregation and deprivation as well as fears, safety and belonging. More exactly, we can query whether Spatialized Islamophobia corresponds to or contrasts with the spaces where Muslims and the main religious buildings are located, or with disadvantaged spaces or even with exclusive spaces. Geographical tensions operate in cities and districts, and reveal the areas that seem to attract or discourage the practice of Islamophobia, as well as phenomena of exclusion/inclusion that seriously undermine the spatial mobility and belonging of victims.

Measuring and mapping Islamophobia are particularly effective at the urban scale and are essential if we are to carefully monitor and fully observe how it functions. The exploration of cities makes it possible to discuss the finer spatial elements of Islamophobia and highlight its processes of operation and distribution in the two capitals of Paris and London. Drawing on the same previous quantitative data, this comparative case study offers an interesting cross-urban analysis of discrimination as well as innovative spatial readings using maps. The geographies of urban Islamophobia show a specific logic of distribution and the importance of particular spaces and places, such as the centres, suburbs, communication axes, pockets of segregation, public areas, transport networks and public institutions, which offer striking differences in terms of where Islamophobia is most likely to occur in each city.

This geographical work initiates discussions about the fears of victims of Islamophobia and their inability to move safely across the city. Henri Lefebvre's (1996) 'right to the city' is premised on the emancipatory struggle for disadvantaged people to reclaim the use of urban spaces and their attachment to everyday places. The daily violence that many Muslims experience in public space restricts their mobility and significantly questions their 'right to the city'. Geographers, in particular, need

DOI: 10.4324/9781003019428-3

to provide spatial answers to how religious discrimination is constituted by using a more qualitative (or mixed-methods) approach. Feelings of spatial belonging are constantly challenged, and victims of Islamophobia can feel safer, therefore more connected, in one area rather than another. The findings for the Parisians and Londoners reveal that victims mainly pursue avoidance strategies when negotiating the city and that their home neighbourhood represents for them a safe place and a symbolic spatial anchorage.

Spatialized Islamophobia in cities and districts, and related social problems

In recent centuries, urban spaces have expanded considerably, and nowadays more than half of the world's population lives in a city (United Nations, 2015). Cities present multiple sets of communities, housing and activities (Grafmeyer & Authier, 2011), and urban geographers have largely shown this plurality, in particular Brian Berry (1964) who explains that cities can be seen 'as systems within systems of cities'. He considers that urban structures are not only different from one city to another but within the same city. A city has indeed many facets varying according to its inhabitants, their social practices and spatial representations. Consequently, it has various socio-spatial forms, and represents above all a space for socialization which can demonstrate various interactions between specific social groups.

These interactions generally reveal social dysfunction (Tissot & Poupeau, 2005), and Spatialized Islamophobia is a good example. The urban space is at the centre of discussions fuelling anti-Muslim rejection. For example, the visibility of markers of Muslimness in many cities of the world (such as the headscarf, the full-face veil, the mosque, the minaret, the call to prayer or street prayer) has caused significant moral panic, rendering the study of urban and infra-urban Islamophobia highly relevant. This is what this chapter will analyse by focusing more locally on the interaction between existing social groups at the scale of cities and districts.

The feminist geographer Gillian Rose (1993) argued that unequal social relations are both expressed and constituted through spatial differentiation, and some human geographers have even mapped, for example, various forms of racism across the city (Forrest & Dunn, 2007, 2010; Nelson & Dunn, 2017). The experience of Islamophobia is documented in a variety of contexts, such as airports (Bennett, 2006; Blackwood, 2015), workplaces (Ghumman et al., 2013), public institutions (Najib, 2020a) and public spaces (Listerborn, 2015, Gökarıksel & Secor, 2012, 2015), and there is scope for further investigation of its prevalence in urban spaces. Urban studies on Islamophobia are more recent, yet are starting to multiply (Najib & Hopkins, 2019, 2020; Najib, 2020a; Itaoui, 2016, 2020; Hancock, 2020). These are usually localized case studies (sometimes comparative studies) focused on important cities around the world and provide a better understanding of how Islamophobia is manifested in metropolitan areas. Global cities undoubtedly present a risk of Islamophobia which can be spatially distributed either in a diffuse or a site-specific way, in connection with other geographies (ethnic, religious, social, economic, etc.).

First, the study of Muslims' residential location reveals important segregated areas within cities (Peach, 2006a; Gale, 2013; Falah & Nagel, 2005; Phillips, 2006). The study of segregation has a long history in urban and social geographies, and in the Muslim populations its extent is not negligible. Some districts of various cities are strongly associated with believers of a particular religion, and several works focus on mapping the religious groups and the patterns of segregation that they form (Valins, 2003; Watson, 2005; Peach, 2006a; Gale, 2013). These works do not apply only to Muslims; many other religious groups show significant gatherings. Here, what is most important to notice is not the high level of Muslim segregation, but rather that the discussions on segregation for Muslims are often negative and close to those denouncing their lack of integration and citizenship.

At the global and national scale, I have already analysed how Muslims can feel excluded from citizenship, and now, at the urban and infra-urban scale, I show how Muslims are negatively portrayed as segregated communities living separate lives. As explained previously, Muslims are part of national society, and this chapter is about demonstrating that they do not tend voluntarily to self-segregate and keep their distance from other groups. Indeed, like integration, segregation is also a mutual process. Negative narratives depicting Muslims as a monolithic bloc operate in the same way to attribute to them a homogeneous will to self-segregate. Many authors have challenged this position of self-segregation and worked to reshape Muslim geographies as spaces of connection and diversity (Phillips, 2006; Aitchison et al., 2007). In itself, the word 'segregation' has negative connotations (Brice, 2009; Peach, 1996; Musterd, 2003), but applied to Muslims, these connotations seem to worsen. Indeed, residential clusters based on other variables do not cause the same level of hostility in the media or with politicians. This is the case with districts specifically promoting the sexual orientation of their residents (e.g. LGBT district) or their education (e.g. university campus), their employment (e.g. artists' district), their wealth (e.g. gated community), their ethnicity (e.g. Chinatown) or even another religion (e.g. Jewish district), all of which are widely regarded as acceptable (Sardar, 2009). The Muslim segregation is rather dismissed as a completely abhorrent phenomenon, and suggests that social problems are stronger there than elsewhere. Therefore, this particular attention to the 'negative' and 'dangerous' segregation of Muslims must be discussed and criticized.

When studying Muslims, religious gatherings are usually associated with ethnicity. The racialization of Islam naturally highlights ethnic segregation analysed through the religious lens, and often leads to associating Muslim segregation with non-White areas. Muslim segregated areas are compared and opposed to White areas, which are never described as segregative but rather as overwhelmingly innocent and positive (Hopkins, 2009). But there is an important phenomenon called 'White flight' (Schelling, 1978; Kruse, 2005; Avila, 2005), showing that White people voluntarily flee areas where non-Whites live.

This phenomenon has been observed for a long time, especially in the 1960s and 1970s following studies on Black/White dualism. For instance, the works of the economist Thomas Schelling allow us to understand how individual preferences lead to homogeneous groupings of people. In a context of Black/White ethnic segregation in the United States, Schelling proposed a model (1969, 1971) that simulates these two distinct populations, initially distributed to present a perfect mix like on a chessboard (even if this scheme of total mix does not exist in any area). Schelling, by slightly disturbing this perfect mix, shows that there is a 'natural' process at work that ultimately leads to a strong segregation. He determined a racial tolerance threshold, explaining that if only a third of similar neighbours are left, then these residents will also want to move. This model reveals an important gap between individual preferences and collective consequences, empirically confirming the validity of the 'White flight' theory. More specifically, Mason (2010) reminds us how Black residents are in favour of living in racially mixed areas, unlike White residents (Charles, 2003), and finds that this 'White flight' cannot be explained by racial socioeconomic differences, as rich Blacks are nearly as segregated from Whites as poor Blacks (Massey, 2004).

Concerning the spatial distribution of Muslims, urban geographers have shown that it often corresponds to the geography of relative material deprivation, particularly in the United Kingdom and France (Falah & Nagel, 2005; Peach, 2006a; Phillips, 2006; Gale, 2013; Najib, 2020a). Thus, geographical studies on Muslims must specifically connect to spatial patterns of poverty. In studies on economic segregation (Jargowsky, 1996; Voas & Williamson, 2000; Wilson, 1987; Massey & Denton, 1993), it is well known that the wealthiest people deliberately distance themselves from the rest of society. Indeed, the many social groups that make up society are not randomly distributed across the city; the distribution is strongly affected by their economic situation and residential behaviours (Najib, 2020b). The housing market is sensitive to economic and political pressure and directly impacts upon people's budget. Often, it is the economically (and politically) weak who suffer the most, and important dynamics of separation become outrageously visible in cities. This separation of peoples and spaces leads to serious inequality in access to employment, education, housing quality and social integration (Wilson, 1987; Jargowsky, 1996; Friedrichs et al., 2003; Gale, 2013), also in the expression of discrimination, racism and Islamophobia (Jackson, 1987; Musterd & Ostendorf, 2005; Wacquant, 1993; Letki, 2008; Bowyer, 2009; Forrest & Dunn, 2007, 2010; Itaoui, 2016, 2020; Najib, 2020a). Consequently, the spatial distribution of Muslims in cities traces an important economic and ethnic segregation that is linked to White people's individual preferences.

This theory of 'White flight' has been clearly observed between Muslims and Whites, particularly in multicultural Britain following the 2001 riots in Bradford, Oldham and Burnley. These cities were perceived to be affected by 'racial' divisions and a lack of local unity (Cantle Report, 2001) and, in this context, Muslims and Whites were seen as two distinct groups leading 'parallel lives'. But only Muslims were blamed for this segregation, even though some had tried to move from Muslim areas to White areas to enjoy better housing, schools and intercultural mix (Phillips, 2006). Once settled, these new Muslim residents expressed serious worries about racist harassment. Many studies acknowledge the existence of Islamophobic incidents in these towns, but first consider that these riots were mainly to do with economic, social and political tensions rather than racial and religious tensions (Sardar, 2009; Massey & Tatla, 2012). Thus, people's attitudes are not only a consequence of spatial context; they can also be a consequence of collective behaviours in place-specific circumstances (Forrest & Dunn, 2010) and political situations (Gökariksel & Mitchell, 2005). In any event, it seems fairly clear that living a separate life in a particular area does not raise any questions if the resident is White, but it does when the resident is Muslim.

Many researchers have already shown in many cities of the world a direct relationship between the degree of residential segregation and the degree of social distance between groups (Taeuber & Taeuber, 1965; Massey & Denton, 1993; Peach, 1996; Voas & Williamson, 2000; Gale, 2013, Najib, 2020b). They explain that this degree of residential segregation is a significant reflection of social separation, regardless of the resident's personal characteristics. Likewise, a Muslim resident is primarily a resident of an area, before being a Muslim; this means that his/her residential choice will use the same criteria as everyone else when looking for new accommodation, such as its attributes (price, size, etc.) and the area's amenities (reputation, green space, proximity to family, schools, workplace, transports, shops, etc.) (Najib, 2013). Like most people, Muslims choose to move house when their family grows, that is to say when they need more room (Brice, 2009).

Their segregation is mainly due to the constraints of the housing market, but also to security and worries about racist and Islamophobic behaviour (Phillips, 2006). Therefore, their final choice is more limited and may focus predominantly on Muslim areas (Hopkins & Smith, 2008). The affinity grouping of peers should not be viewed systematically as negative, in particular for religious and ethnic-minority groups, since this grouping facilitates social networks and ties of solidarity favourable to their professional and socio-political integration (Peach, 1996; Charmes, 2009). Thus, Muslim populations will first choose the type of accommodation that they need and can afford, ideally located in a neighbourhood that offers interesting general amenities, including religious amenities (such as proximity to mosques, Islamic schools, *halal* shops, etc.). At the end of the day, these limitations certainly make it more difficult for Muslims to live in non-segregated areas.

Religious amenities are important to minority religious groups, whether they are Muslims, Jews or Sikhs. Many studies have demonstrated the role of religious groups in shaping the built and physical environment of cities. The urban landscape is strongly affected by religious architecture (places of worship, cultural buildings, pilgrimage routes, etc.), and many studies focus on mosques, notably in the United Kingdom (Naylor & Ryan, 2002; Gale, 2009; Peach and Gale, 2003), in Australia (Dunn, 2005), in France (Gartet & Id Yassine, 2013) and in the United States (Kahera et al., 2009; Kahera, 2010) that all report significant conflictual and racist repercussions due to their construction. Here, the discussions stem mainly from concerns about urban design and planning that refer to institutionalized conceptions of the religious space, but also connect significantly to debates on multiculturalism and citizenship. Such discussions can socially encourage negative public attitudes towards Muslims and Islam (Allen, 2019) and politically shape the analytical frame of religious space policy with national and cultural reorientations (Kong, 2009; Gale, 2009). The religious space is indeed governed by the State and local policies that grant equality for all religious groups to have adequate places of worship, while at the same time controlling their visible public expression (size of mosques, height of minarets or audibility of calls to prayer). Religious identity can also be expressed through the form of other religious buildings such as schools (Dwyer & Meyer, 1995; Merry, 2005; Archer, 2009), *halal* shops and restaurants (Riaz & Chaudry, 2003; Isakjee & Carroll, 2019), and these have also been the subject of political debates concerning integration and segregation. Minority religious groups obviously have the right to a range of community services, but again these requirements seem more problematic and less accepted for Muslims than for other religious groups (Newman, 1985).

It is therefore necessary to think about an urban system and more precisely a territorial organization of districts that minimizes the stigmatization of Muslims. But the problem is that the socio-spatial organization of cities cannot be achieved without a thorough understanding of the whole city system (Harvey, 1973). Political decisions are therefore difficult to define and actions difficult to carry out locally, even when they are motivated by important social objectives such as the fight against Islamophobia. The social and spatial forms of urban systems must be better identified in order to anticipate the various increasingly complex interactions. In fact, in the city everything affects everything else: people are connected to each other and organizations are linked to people, just like housing, transport and job opportunities (ibid.). The question of controlling the urban and infra-urban organization to eradicate the problem of Islamophobia is crucial, and this entire book seeks to provide answers to counter the problem of Spatialized Islamophobia and reduce it at each spatial scale (from the globe to the body (and mind)). Therefore, research on geographies of Islamophobia represents an important development in public engagement, education and policy-making. Indeed, knowing more about the spatial dimension of Islamophobia (which highlights specific places, people and functioning) can help decision-makers better define an appropriate local policy to address this anti-Muslim issue. Finally, this first section highlights, in a theoretical way, the existing social problems involving Muslim populations that are observable at the urban and infra-urban scale. Thus, linking these urban geographies to Spatialized Islamophobia will yield significant insights into its quantitative functioning through exploring the specific examples of Paris and London.

Mapping urban Islamophobia in Paris and London

Urban models of Islamophobia in Paris and London

The mapping of urban Islamophobia represents an important innovative contribution of research in social and urban geography, and shows the types of spaces in which anti-Muslim acts take place. The map is the geographer's favourite tool and sometimes even constitutes the outcome of his/her research. Here, the maps describing the spatial distribution of Islamophobia in Paris and London show the spatial logics specific to each of the studied cities, and required the collection of quantitative geocoded data.[1]

Paris and London are important European capitals affected by multiple phenomena of interest (urban macrocephaly, cultural and religious diversity, Muslim overrepresentation, Islamophobic risk, etc.). This is why it is interesting to compare them, in particular their urban patterns of Islamophobia. In Geography, urban models are generally represented by specific forms (such as the point, line and area), and in urban sociology they generally refer to three well-known models that display the urban elements of chief importance (such as the centre, transport axes and focal points). More exactly, these models stem from research on urban inequalities carried out by the Chicago School. The first is the concentric model, also called the centre-periphery urban model (Burgess, 1925), contrasting the city centre with its suburbs. The second is the sector model which takes into account the importance of specific axes, and notably transport axes (Hoyt, 1939). The third is the multiple nuclei model (Harris & Ullman, 1945) which describes an urban mosaic structure in which focal points present a certain competition between the primary and secondary centralities.

Figure 3.1, comparing Greater Paris and Greater London,[2] is from Najib and Hopkins' (2020) more detailed interpretation of the various quantitative data, and readers can refer to this article for further discussions. What I wish to reiterate here is that the Parisian centre is the most affected *département* of Greater Paris, while in London, Islamophobia shows no differences between Inner London and Outer

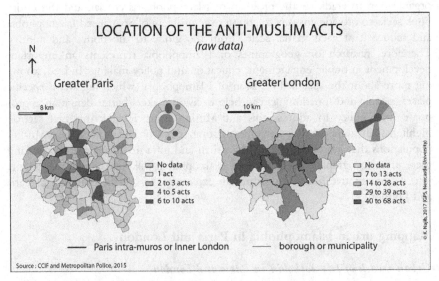

Figure 3.1 Spatial distribution of anti-Muslim acts in Paris and London
Note: The names and locations of the various zones are given in Appendix 1.

London and instead displays important horizontal and vertical lines. Indeed in Paris, anti-Muslim acts take place more often in the centre and gradually decrease away from this prestigious centre. Other isolated and scattered focal points are underlined in more or less dark tones, notably in the northeast of Central Paris and in the south. In London, Islamophobia seems to be more spatially diffuse (because all boroughs are affected, and the acts are more numerous) and seems to extend mainly along the major roads of London. For example, a dark horizontal line, showing a worrying level of Islamophobic acts, runs just north of the Thames from Newham to Hounslow, including boroughs known for their large Muslim populations. Consequently, it can be said that Islamophobia in Paris describes a centre-periphery model contrasting the city centre with its suburbs (or *banlieues*),[3] unlike London describing a sector model whose configuration takes into account transport axes. To be precise, spaces of Islamophobia rather correspond to places of Islamophobia, that is to say the exact places where anti-Muslim acts take place (Najib, 2020a). In Paris, the majority of Islamophobic acts occur in public institutions (like town halls and training centres) (CCIF, 2016); which can explain this important centrality. Meanwhile in London, anti-Muslim acts occur more often in public areas and on public transport (Tell MAMA, 2016); which can reveal significant axes (Najib & Hopkins, 2020).

In summary, this comparison shows that the urban models of Islamophobia differ between cities. These variations are evidence of the spatial nature of Islamophobia and immediately reveal, through the statistical data and cartographic analyses, the presence of various interconnected segregations. The maps and urban models in Figure 3.1 show several divisions between areas: central or peripheral; northern or southern; prestigious or deprived; Muslim or non-Muslim, etc. Segregation is, in fact, a 'statistical concept' (Carling, 2008) describing the uneven distribution of a variable. In this example of Islamophobia, a different and unequal spatial logic is observed in Paris and in London, and specific areas can be regarded as segregated. Indeed, the Paris map already shows that the spaces of Islamophobia are the more privileged and central areas (such as *Paris intra-muros*) that contrast with the situation in the suburbs; and the London map already highlights that Islamophobia occurs in the predominantly Muslim boroughs of Tower Hamlets and Newham (Githens-Mazer & Lambert, 2010; Naqshbandi, 2006). It is therefore important to verify these initial observations via a more in-depth statistical analysis using additional quantitative data. Below, the geographies of Islamophobia are compared statistically 1) to the geographies of socioeconomic segregation in Paris and 2) to the geographies of ethnic-religious segregation in London, according to the availability of data in the two countries of France and the United Kingdom.

Urban Islamophobia in Paris and socioeconomic segregation

In France, no data are available on ethnicity, race and religion. Therefore, in the case of Paris, connecting the geographies of Islamophobia to the geographies of

Muslims and Islam is challenging. Only the connection with socioeconomic geography can be analysed statistically. This has already been explored in depth in Najib (2020a), but the important findings are restated here. The spatial distribution of various socioeconomic and demographic variables was analysed and synthesized on a typological map displaying the significant forms, relations and diversities in Greater Paris. These variables[4] are from the population census and household taxation in 2012 and refer to citizenship, family status, socio-occupational category, qualification, unemployment and income. Figure 3.2 measures and compares the degree of connection between the map presenting the spaces of Islamophobia and the map exposing the socio-spatial divisions of Greater Paris.

French studies have essentially explored the socio-spatial divisions with socioeconomic and demographic variables (Musterd & De Winter, 1998; Tovar, 2011), and the French capital is a prime site for testing these problems of inequality and diversity. Paris is indeed the study area for most research on gentrification (Clerval, 2016), gated communities (Le Goix, 2006; Charmes, 2004), the upper classes and elitism (Pinçon & Pinçon-Charlot, 2014) and also deprived and troubled areas (Wacquant, 1993; Chignier-Riboulon, 2010). It is also the preferred study area for research on territorial and religious discrimination, such as Islamophobia (Hancock, 2013, 2017; Najib & Hopkins, 2019, 2020; Najib, 2019, 2020a) although these are less extensive.

The social geography of Greater Paris is therefore well documented, and the map in Figure 3.2 reveals, thanks to Hierarchical Cluster Analysis,[5] five distinct and contrasting areas: 1) privileged Central Paris (in green); 2) modest suburbs (in orange); 3) privileged suburbs (in blue); 4) deprived suburbs (in purple); and 5) troubled areas (in red) (see Najib, 2020a for a better description of the colours). The exposure of such social divisions testifies to a high rate of Islamophobic acts in the privileged Parisian centre. Islamophobia also affects, to a lesser extent, the deprived and troubled parts of the suburbs, particularly the municipalities of Saint-Denis, Aubervilliers

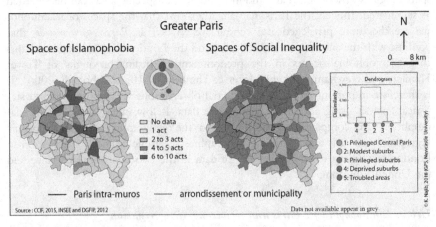

Figure 3.2 Comparison of Parisian spaces of Islamophobia and spaces of social inequality
Note: The names and locations of the various zones are given in Appendix 1.

and Aulnay-sous-Bois located in the *département* of Seine-Saint-Denis in the north-east of Greater Paris. Other focal points are located in the southern suburbs in the municipalities of Créteil and Orly, which are respectively parts of the modest and deprived suburbs. Créteil is better known as an administrative and academic municipality famous for its higher education population, and Orly for its airport. Thus, this statistical and cartographic analysis confirms that the geographies of Islamophobia correspond more to the prestigious Parisian centre and less to degraded areas. Islamophobia intersects with the meanings ascribed to space by various social groups, and highlights well-known phenomena such as socioeconomic segregation (gentrified centre, deprived areas, etc.).

Islamophobia can also supply the residential location of victims (and Muslims in general). France is an important study area for Muslims and their everyday lives because it has the greatest Muslim presence in Europe, both in real and perceived terms (Duncan, 2016). This strong presence is mainly due to the African countries (North and Sub-Saharan) that were French colonies (Hancock, 2009; Laurence & Vaïsse, 2007; Simon & Tiberj, 2013). Algerians and other North Africans as well as Sub-Saharan Africans have migrated to France and been located in areas with a high concentration of poverty, social housing, minority groups, violence, ghettoization and despair (Cesari, 2005; Selod, 2005; Pan Ké Shon, 2007; Preteceille, 2009; Safi, 2009; McAvay & Safi, 2018). These segregated areas[6] where the majority of Muslim populations may live do not necessarily reveal a high rate of Islamophobic acts (Najib, 2019, 2020a). This non-statistical observation is valid; nevertheless, due to the unavailability of ethnic and religious data, it must be considered with caution. This is why, with reference to this observation, statistical data on the residential location of Islamophobia's victims are used.

The collected quantitative data indeed include victims' residential location, and similarly reveal that the spaces of Islamophobia are generally not where the victims live. For example, of all the anti-Muslim acts that took place in Greater Paris in 2015, 65% involved victims who lived in the suburbs, compared to 18% in *Paris intra-muros*. Thus, the victims experienced Islamophobia beyond their everyday spaces, especially in the Parisian centre. More precisely, of all the anti-Muslim acts perpetrated within *Paris intra-muros*, 56% concerned victims living in the suburbs, compared to 39% living in *Paris intra-muros* (including 16% from an *arrondissement* other than where the act took place) (Najib, 2020a). Be that as it may, French Islamophobia primarily represents a specific type of hate crime that relates to institutions and clearly contrasts with British Islamophobia which involves more public spaces.

Urban Islamophobia in London and ethnic-religious segregation

As previously shown, Islamophobia in London mostly takes place in streets, parks, shops, buses, the Tube, and so on (Tell MAMA, 2016; Najib & Hopkins, 2020), making the study of interactions between Muslims and non-Muslims in public spaces even more relevant (particularly because data on race and religion are freely accessible in the United Kingdom). The way in which the different communities interact with one another has always provoked lively debate, especially on existing racial tensions between Muslims and Whites. Research has shown how Muslims can see White areas as risky, unsafe and racist (Archer, 2009), and how Whites can similarly refer to Muslim areas in negative terms of fear and rejection (Phillips, 2006). But, as previously explained, these discussions mainly emphasize and criticize Muslims for their alleged desire to live separately (Cantle Report, 2001; Dwyer & Uberoi, 2009; Phillips, 2006).

In the United Kingdom, ethnic and religious segregated areas began to be mapped statistically in the early 2000s (from the 2001 population census). Peach (2006b) was one of the first to claim that London Muslims together are much less segregated than Sikhs, Jews or Hindus, and that there is a considerable residential separation between Muslims of differing ethnic origins. Likewise, Phillips (2006) was able to explain that Muslim segregation is not self-imposed or desired but is due rather to a racialization of space. Therefore, Muslim segregation is much more complicated (Brice, 2009; Hopkins & Smith, 2008; Gale, 2013) than is assumed. It can be linked to particular ethnic segregation within Muslim populations, but also to a strong economic segregation. Indeed, the presence of Muslims in the United Kingdom can be traced to their South Asian heritage and is the result of waves of economic migration (Peach, 2005; Gale & Hopkins, 2009). Low income and unemployment are disproportionately high among Muslim minorities in Britain who tend to live in relatively segregated areas with high levels of deprivation (Peach, 2006a; Bowlby & Lloyd-Evans, 2009; Sardar, 2009; Abbas, 2009). For example, in the London borough of Tower Hamlets, studies have shown that Muslims are not only affected by poverty, housing problems, poor education and leisure, and health problems, but also by racism and Islamophobia (Datta, 2009; Pollard et al., 2009; Sardar, 2009; Najib & Hopkins, 2020). In this context, it is relevant to measure the extent to which spaces are racialized and to compare them to the geographies of Islamophobia in Greater London.

Strikingly, some areas of London are racialized, and the maps show how divided these spaces are. Figure 3.3 clearly shows the opposition between the spaces where Muslim populations and White people live. For example, directly east of the City of London there is a large proportion of Muslims and a low proportion of Whites in the central boroughs of Newham and Tower Hamlets. Conversely, in the extreme eastern and southwestern suburbs (respectively in Havering, Bexley, and Bromley, and in Richmond upon Thames, Kingston upon Thames, and Sutton), there is a high proportion of White residents and a low proportion of Muslim residents. These areas contrast strongly: the spaces where there is a high proportion of Whites contrast completely with the spaces where the great majority of residents are Muslims, and *vice versa*. The spaces in paler tones where Muslims are underrepresented correspond almost perfectly to those in darker tones where Whites are overrepresented. These observations strongly highlight segregated areas, reminding us of huge urban ghettos and also the phenomenon of 'White flight' that undoubtedly occurs from these areas. Indeed, in the United Kingdom, research on 'ghettos' first took the form of an academic debate on the scale of measurement of segregation of ethnic and religious minorities (Jones & McEvoy,

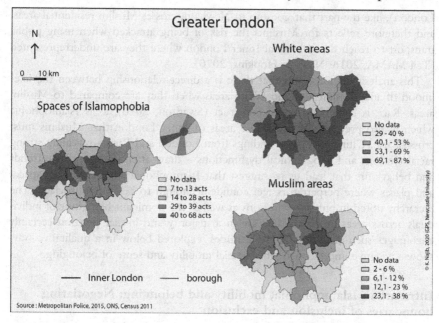

Figure 3.3 London's spaces of Islamophobia and Muslim and White boroughs
Note: The names and locations of the various zones are given in Appendix 1.

1979; Lee, 1978; Peach, 1996) dividing minority groups from White identities (Watt, 1998; Bonnett, 1996).

More exactly, Figure 3.3 demonstrates that the spatialization of urban Islamophobia in London occurs not only in predominantly Muslim areas, but also in some predominantly White boroughs that Muslims may use. Specifically, the dark horizontal line recording a high level of Islamophobia, mentioned previously, highlights different areas. For example, in Inner London, a first important cluster is in the east, notably in Newham, Tower Hamlets, Hackney and Islington which record a high proportion of Muslim residents and a low proportion of White residents (with the exception of Islington). These areas are strongly affected because many Muslims and Islamic buildings are frequently targeted.

In addition, a second cluster is also observed in the western part of Inner London, in the wealthy boroughs of Hammersmith & Fulham and Westminster which record a significant level of anti-Muslim acts but where Muslim residents are less prevalent than White. This cluster rather represents a transit zone; the anti-Muslim acts recorded here occur mainly in areas with good access to public transport (multiple bus stops, Tube stations, etc.) and with high levels of pedestrian activity (such as busy streets and major shopping and tourist areas) (Tell MAMA, 2016). In this sense, the cluster extends in the west to Outer London, to the three boroughs of Brent, Ealing and Hounslow where the great majority of residents are Muslim and where White residents are underrepresented. In the end, this part of the horizontal line (from Hounslow to the influential Central

London), like the part that goes up to Newham, crosses Muslim residential areas, and therefore reflects for Muslims the risk of being attacked when using public transport to reach the influential Inner London where they are underrepresented (Tell MAMA, 2016; Najib & Hopkins, 2020).

This analysis demonstrates that there is a direct relationship between the likelihood of anti-Muslim acts and White areas when they are compared to Muslim areas. Race is certainly a social construct (Swanton, 2016), as is Islamophobia which is also prevalent in the White areas of Inner London that Muslims must cross or use. Ultimately, these findings from London and Paris – revealing strong racial tensions and geographical dysfunctions – draw attention to specific trends and behaviours that lead us to suggest that Islamophobia arises mainly in spaces and places where perpetrators feel confident enough to behave antisocially. The hierarchy noted in public institutions as well as the dominant behaviour of individuals across areas in which they are in a majority and insider position certainly encourages such attitudes. These attitudes, explored below in a qualitative way, show a negative impact on victims' spatial mobility and sense of belonging.

Infra-urban Islamophobia, mobility and belonging: Negotiating boundaries of inclusion and exclusion

The link between divided spaces and avoidances strategies has been discussed (Boal, 1969; Hwang & Murdock, 1998; Bader & Krysan, 2015), and shows how people with diverse identities and senses of belonging share and negotiate urban spaces. In the city, there are undoubtedly areas where Muslims do not feel welcome or safe. Their ability to move safely around different districts of the city is affected, and demonstrates complex mapping systems of where they feel they can go and can bond. Tellingly, their sense of insecurity restricts their mobility, directly influencing their sense of belonging to specific spaces and places. Facing everyday Islamophobia, Muslims feel the need to negotiate boundaries of 'inclusion' and 'exclusion', mirroring boundaries of 'self' and 'other' (Salih, 2000). This can only be explored using qualitative methods (mental maps, commented walks, surveys, focus groups, individual interviews, etc.).

The perceived geographies of Islamophobia undoubtedly connect to geographies of exclusion, no-go areas, safety, fear and belonging; these are logically examined at the infra-urban scale. Geographical debates on the exclusionary dimensions of urban districts have already discussed the contested ways in which Muslims negotiate and perceive cities, such as in Istanbul in Turkey (Gökariksel, 2012), Malmö in Sweden (Listerborn, 2015), Paris in France (Najib & Hopkins, 2019), Sydney in Australia (Itaoui, 2016) and San Francisco in the United States (Itaoui, 2020). Consequently, Islamophobia is also spatially constructed, and different anxieties can be mentally mapped to specific urban districts. Oppositions and hostilities against Muslim visibility have already manifested themselves in urban districts (against mosques, schools, shops and even against Muslim men and women), and include a diversity of discrimination, degradation, harassment, intimidation, verbal aggressions and physical assaults that result in potentially life-threatening experiences (Chao, 2015; Samari, 2016). Muslims have therefore developed spatial strategies by avoiding the spaces and places where they know in advance that they will not be welcome or accepted (Garner & Selod, 2015; Gökariksel & Secor, 2015; Hancock, 2013; Listerborn, 2015; Najib & Hopkins, 2019). This type of self-censorship draws a particular 'geography of risk' (Noble & Poynting, 2010; Tell MAMA, 2016) that is perceived and experienced by Muslims who enter into a negotiation process that promotes, above all, their invisibility through avoidance.

In this context, Muslims' concerns regarding their everyday security and their potential exposure to Islamophobia lead them to think carefully about the means of accessing, engaging and moving across the city's districts. Experiences of oppression generally limit mobility (Farris, 2017; Hanson, 2010; Kwan, 1999; Rose, 1993) and, in the case of Islamophobia, it is particularly veiled Muslim women who avoid specific districts that are unfamiliar to them or that represent possible anti-Muslim hostility (Kwan, 2008; Listerborn, 2015; Gökariksel & Secor, 2015; Najib & Hopkins, 2019). They rarely deviate from their usual activities and spaces, as for them these are 'safe spaces' where they can enjoy time with their friends, neighbours or relatives, and escape the potential risks of other contexts (Kwan, 2008; Bayoumi, 2010). Some spaces are classified as so dangerous and unwelcoming to veiled Muslim women that they never set foot there alone, describing them as 'no-go areas' (Tell MAMA, 2016; Zempi & Chakraborti, 2014; Carr, 2016).

This spatial exclusion is embedded in everyday geographies and questions the very presence of veiled Muslim women, or more precisely their right to be visible and to access certain parts of the city. The 'right to appear' (Butler, 2011a) and, especially, the 'right to the city' (Lefebvre, 1996) generally reveal tensions between peripheries and centres, specifically an inability to access the prestigious centre of most cities. This right to centrality was particularly explored by bell hooks (1990, 2008), who explains how the southern Black community is segregated from the White city centre, and how Black Americans feel out of place in this non-inclusive centre. It was also discussed by Iris Marion Young (1990) in her feminist study on gentrification where she shows, in connection with the concept of 'unassimilated otherness', how the attraction of the centre for women breaking out from common appearances can be a challenge.

Centrality is definitely a concept that matters in terms of social, political and cultural effects. Harvey (1973), in this sense, reminds us how cultural attitudes in the inner city have always been different from those in the suburbs. These differences may also apply to Muslims who generally live in enclaved and marginal peripheries (Peach, 2006a; Bowlby & Lloyd-Evans, 2009; Datta, 2009; Pollard et al., 2009; Najib & Hopkins, 2019; Najib, 2020a). Muslims can feel excluded from the centre because of their social class and residential location, but also because of their religious appearances and cultural practices. This feeling of exclusion certainly establishes sharp boundaries between the centre and the suburbs, between safe and unsafe spaces, and between familiar and unfamiliar spaces that lead to a collective damaged and biased sense of belonging in certain geographies.

A lack of spatial belonging underpins the subtle exclusion and marginalization of people who are deemed to be out of place (Holloway & Hubbard, 2001). Individuals are generally included into many spatial scales of belonging, and this multi-scalar reality – referring to the many facets that make up the spatial identity of the victims – shows an important relationship with feelings of well-being, comfort, safety and inclusion. Depending on their various circumstances of discrimination, victims of Islamophobia feel safer at the finer familiar spatial scale, challenging their other sense of belonging to a larger entity.

For example, some studies have already shown that Muslim populations living in Copenhagen, Denmark, or in Marseille, France, feel more connected to their city than to their country (Koefoed & Simonsen, 2010; Lorcerie & Geisser, 2011), while others have emphasized an important connection with their district of residence (Kwan, 2008; Githens-Mazer &Lambert, 2010; Zempi & Chakraborti, 2015; Najib & Hopkins, 2019). This preference for local belonging is conditioned by the climate of suspicion that hinders Muslims' access to all spatial resources (Amin, 2002; Phillips, 2006). Their spatial attachment is naturally rooted in more limited, familiar and reassuring spaces which also represent the marginal spaces of society. Muslim minority groups are, indeed, defined by the gaze of the majority (the dominant), which necessarily impacts upon their sense of belonging as well as the construction and manifestation of their Muslim identity.

Urban encounters are shaped by these types of societal attitudes, but also by political discourses, structural inequalities and imaginary geographies (De Certeau, 1984; Amin, 2002; Andersson et al., 2011). Wilson & Darling (2016) show the significance of encounters to the politics of the city by explaining that the city is not simply a space of interactions, but rather represents a battleground of political relations, pushing apart different groups of people who constitute their own identities and make their own claims to spatial visibility (Isin, 2002, 2007; Allen & Cochrane, 2014). In this context, urban fractures seem to be maintained rather than challenged, and urban togetherness does not yet seem to be on the move. Finally, questioning the zones of contact and encounter with either a majority or a minority, as well as the spatial positionality as either an insider or an outsider, relates spaces to opposing functions (inside/outside spaces, inclusive/exclusive spaces, attractive/repulsive spaces, etc).

These are key issues reflecting the diversity of contemporary urban life and the nature of spaces that explore the way in which identities are formed, reproduced and marked off by one another. Indeed, there are many other identities targeted by discrimination in the city's districts, and they too develop their own spatial strategies and sense of belonging to avoid any hatred situation. Thus, social and urban geographers must continue to provide valuable insights into the impact of discrimination (particularly Islamophobia) on the spatial exclusion and marginalization of victims in order to better understand why a person – living in a nation-state that upholds law, equality, fraternity and freedom – considers at some point that such a space is too difficult to access or to belong to.

Pushed to the margins: Reflections of Parisians and Londoners

Negotiating mobility and fear in urban streets

In the study of racism and discrimination, place matters (Forrest & Dunn, 2007; Najib, 2020a); that is, the place where victims and perpetrators encounter each other, where they work or move across the street. Encounters are about difference (Wilson & Darling, 2016): we can therefore understand Henri Lefebvre when he explains that social encounters make the urban (and I would add the infra-urban) a place of 'permanent disequilibrium' (1996: 129, as cited in Wilson, 2016). More specifically, urban streets are the terrain of encounters, as well as sites of domination, resistance and anxiety (Fyfe, 1998; Valentine, 2008; Amin, 2012), and for victims of Islamophobia, as soon as there is any form of commuting and mobility, there is an increased risk of discrimination and aggression (Najib, 2020a).

To better explore their ways of negotiating urban streets, this section analyses the actual lived experiences of victims of Islamophobia living in Paris and London, by contrast to the theoretical and quantitative analyses of Islamophobia presented earlier in this book. I conducted qualitative interviews with Parisians and Londoners in the summer of 2017 and winter of 2018 (33 victims in Paris and 27 in London) whom I had met either at the CCIF and MEND's offices or in a public place (café, park, etc.). All the interviews were audio-recorded and lasted between 28 and 106 minutes (60 hours in total). From their individual narratives,[7] I collected important data on how they use spaces, how they avoid certain streets and places, and how they change their habits and practices in relation to their mobility. The two samples do not necessarily represent all Muslim populations of Paris and London. Instead, they correspond to the typical profile of a victim of Islamophobia, or at least a victim who can report their experience of discrimination.

The quantitative data correspond more or less to the qualitative results of the sample, since the typical victim interviewed in either Paris or London is a young *hijabi* woman from a foreign ethnic background (North African in Paris; South Asian in London). The samples describe quite similar findings from Paris and London, other than that women who wear the *niqab* live exclusively in London (as expected, because in France the full-face veil has been banned in public since 2010). Both categories of respondents are 'highly educated', but there are fewer executives and professionals among the French (21%) than the British (48%). The unemployment rate is also higher among those interviewed in Paris (36%) than in London (26%). All the respondents have French or British nationality, with the exception of two women in Paris who have North African nationality and two women and one man in London who have another European nationality and an African heritage.

All the interviewees consider themselves Muslims and religious (apart from one woman living in Paris who does not give a precise answer about her religious practice). This high level of religiosity can be explained by the fact that the majority of victims of Islamophobia are visible Muslims who obviously consider themselves to be mainly religious. Besides, the majority of respondents live in areas renowned for their Muslim populations: 82% of the French interviewees live in the first suburban ring of Paris (mostly in the *département* of Seine-Saint-Denis), and 67% of the British interviewees in Outer London (mostly in Waltham Forrest and Redbridge in the northeast of Greater London and in Newham in Inner London). Most of the interviewees have lived in their city for more than 10 years; they are generally familiar with the Paris and London regions and thus are more able to share their experiences of negotiating their mobility, habits and fears in the city streets.

The following analyses focus tightly on veiled Muslim women (26 in Paris and 20 in London), since they are the main victims of Islamophobia and are primarily worried about spatial mobility. Muslim women who do not wear the headscarf do not have the same concerns and fears, and certainly not the same mobility strategy. Muslim men in general do not have serious spatial limitations comparable to women. Indeed, urban mobility is primarily circumscribed by gender, and women are more frequently exposed to verbal and gestural assaults in urban streets than men (Bondi & Rose, 2003).

First of all in Paris, as already published in Najib & Hopkins (2019), veiled Muslim women consciously or unconsciously avoid the Parisian centre. Some explain that they have not been to Central Paris for a long time because they have nothing to do there – even though *Paris intra-muros* offers rich and diverse cultural activities, places of entertainment, parks, shops, and so on. Other women clearly explain that they do not visit the Parisian centre because they feel out of place. They feel uneasy and more stigmatized there than in the suburbs, as several interviewees explain, in particular Kenza (a 30-year-old French-Moroccan woman in *jilbab*):[8]

> Not at all, I do not go anymore to Central Paris, not at all, at all, at all, as soon as I can avoid Paris, I avoid it. I avoid Paris because I don't feel at all in my place. I won't go there because of the staring eyes, because of the comments, there, I am really clearly afraid of being attacked in Paris, I will not venture there. (Kenza)

In suburban areas, they feel more accepted and more 'part of the landscape', in the words of some participants. Camélia (a 30-year-old French-Algerian woman in *jilbab*) adds, in this context, that she feels more comfortable in the suburbs because she sees the Parisian centre as more of an unsafe space:

> I have the feeling that in the suburbs, people react faster in the sense that they are ready to defend you. For example, if you go to a 'cité' [a troubled urban area of the suburbs (see endnotes 3 and 6)], you will never see a veiled woman being attacked without a young man reacting. It's impossible! Never in life! Whereas in Paris, it's more complicated... I don't know, if you are walking around V. Street [a certain Paris street], for example, I'm not sure there is someone who intervenes. (Camélia)

Here, the threat of a potentially unpleasant welcome or Islamophobic incident calls into question the right of veiled Muslim women to access the Parisian centre. Clear geographical tensions between *Paris intra-muros* and its *banlieues* are uncovered, describing oppositions and contradictions between spaces generally well considered and socially valued but where the victims of Islamophobia feel vulnerable, with spaces generally stigmatized and feared by the majority but where these populations feel comfortable, as several studies have shown (hooks, 1990, 2008; Young, 1990; Harvey, 1973; Listerborn, 2015; Hancock, 2017; Göle, 2003; Najib & Hopkins, 2019).

But Camélia specifies that she does not feel comfortable in all parts of the suburbs, singling out the *département* of Hauts-de-Seine ('92' is the *département* zip code) in the western suburbs, particularly the towns of Boulogne-Billancourt and Neuilly. This part of Greater Paris is wealthier (see Figure 3.2) and does not attract veiled Muslim women, as explained by Samia (a 28-year-old French-Moroccan woman in *hijab*): '*I don't go to the chic 92*'; or Suzanne (a 31-year-old French convert in *jilbab*): '*If you go in [suburban] areas that are a little bit more chic, I think that's it; people would be a bit more aggressive or challenged by the headscarf*'. Therefore, the geographical tensions between the centre and the suburbs seem to be reduced when we zoom in on certain areas and take into account their level of wealth. The interviewees rarely visit the wealthy districts and streets of both the suburbs and *Paris intra-muros*. However, on the wealthiest and more luxurious streets of Paris, some veiled Muslim women feel totally at ease and describe them as 'very welcoming' compared to the privileged *arrondissements* of Paris (e.g. the 16th *arrondissement*). They report that they are not noticed in the famous Avenue of the Champs Elysées, because it is assumed that they are foreigners from wealthy Middle Eastern countries, as Myriam (a 31-year-old French-Algerian woman in *hijab*) details:

> If we wear big black sunglasses, it's OK; they can take us for women from Dubai or Saudi Arabia, we can pass!... The very tourist places, it's OK; they are used to it, they see a lot of people passing, it's OK! All places such as Champs Elysées, there are no worries. On the other hand, if you go to spaces such as the 16th or the 15th arrondissements, as soon as it is less tourist spots, we have some insults and remarks. (Myriam)

Some interviewees enjoy visiting this famous Avenue, but the majority of them nevertheless explain that they avoid crowded streets and places, especially when alone or with young children. For instance, Janna (a 31-year-old French convert who wears the *hijab* or *jilbab*) describes her spatial limitations after suffering a very serious physical assault:

> Yes my activities are limited and timed. I avoid Central Paris, except when I am with my husband. I avoid crowded places where they are many people. Mingling with the crowd, it's not my cup of tea at all, it's over for me. (Janna)

As a consequence, French veiled women are extremely careful in choosing a place to meet their family or friends. They feel welcome both in luxurious places, such as five-star restaurants and cafés (if they can afford them) and, conversely, in inexpensive community places and major well-known chains (such as McDonalds or Starbucks). They never use random places; instead, they use word-of-mouth or other recommendations (sometimes using blogs on the internet) or already known and used places. In the end, these testimonies underline the feeling of insecurity among veiled Muslim women, limiting their spatial mobility and accentuating their marginalization and exclusion from certain streets and places. With increased intensity of Islamophobic acts, some visible women start to feel unsafe even in places where they are insiders, such as mosques for example, albeit without stopping visiting them. The women interviewed are particularly afraid for their husbands, since men are more likely than women to go to the mosque. Here, Sabrine (a 30-year-old French-Algerian woman in *hijab*) explains:

> *I think about mosque also generally after what happened in Quebec City. So here we are, we think about it, it's in a corner of our head. My husband when he goes to the mosque, I'm not reassured. (Sabrine)*

In London, veiled Muslim women do not avoid privileged or central districts compared to the Parisian victims. They rather avoid non-diverse areas, especially predominantly White areas. They are less comfortable in these areas and some even mention being scared. This is the case for Meena (an 18-year-old British-Somali girl wearing a turban) who explains:

> *If I'm going out and I'm in a predominantly White area, I get scared, not scared but you have that fear inside you; is that person not going to treat me the same? [...] If I were to go to workplaces such as Westminster, it's predominantly White, and I wouldn't feel safe. (Meena)*

As a result, veiled Muslim women feel more comfortable in diverse areas (whether with Sikhs, Hindus or Blacks) and particularly in Muslim-majority areas. They avoid, specifically, the streets and places that Muslims rarely use such as high-street pubs and bars. They identify this type of street and place as unsafe and 'scary', as Anila (a 40-year-old British-Pakistani woman in *hijab*) explains:

> *Outside the pubs on Saturday nights, it really scares me because I just walked on this high road once or twice, even if they're not targeting you, but they make noises if they see a person with a hijab walking around. Even at 8 or 9 pm. So, I avoid that. (Anila)*

Just like other studies have shown (hooks, 1990, 2008; Young, 1990), Anila fears the 'Othering' gaze. This is particularly the case for Yusra (a 23-year-old British-Pakistani woman in *niqab*), who explains how people can stare at her while walking past a pub: '*I have had them looking in a weird way at me, or saying something like, "Look, look, look" that kind of thing'.* Veiled Muslim women, and particularly women who wear the *niqab*, explain how they struggle to negotiate everyday Islamophobia in urban streets and on public transport. Indeed, the majority avoid public transport in order to minimize any Islamophobic risk, but also in response to geopolitical events. This is the case for Nadia (a 35-year-old British-South Asian woman wearing the *hijab* or *jilbab*) who has experienced many Islamophobic incidents on public transport, particularly following the 7/7 London bombings in 2005 and the Brexit referendum results in 2016. Consequently, Nadia avoids public transport as much as possible:

> *Yes I would drive in rather than use public transport, and I would basically be limiting how much I need to walk… With public transport I'm slightly wary of… I'm slightly wary where I stand on the [Metro station] platform. I will try to make sure that I'm not so close so that anyone can try to push me. (Nadia)*

When reading Nadia's quote, we can easily understand why *hijabi* and *niqabi* women mainly avoid public transport in London. Islamophobic incidents are very often experienced on public transport in the United Kingdom to the point that they have almost become the norm, as described by Latifa (a British-South Asian woman in her thirties wearing a *niqab*) who was verbally abused on the bus by a man calling her an ISIS terrorist, with no reaction from the other passengers. She laments:

> *People are so immune to racism, hatred, hate speech, Islamophobic comments, it's become such a norm that it doesn't make anybody raise their eyebrows to condemn and frown at a behaviour like that. It's so ordinarily acceptable. (Latifa)*

This time, Latifa did not want to let this incident pass, and decided to react firmly and publicly to report it to the police in order to educate people and show them that Islamophobia is not right. She explains:

> *So, I became quite loud and then I decided… I do not exaggerate the situation but I do make a big deal out of it in terms of, for me it's making people in the public know what is happening is wrong, that we're not going to tolerate it, that we're not terrorists. (Latifa)*

When the police boarded the bus, the passengers began to show some interest because they were upset that the bus had been stopped. Latifa felt the need to stand up in this situation by reminding people, *'Hey, there's injustice going on here, there's racism here'*. These observations not only demonstrate how veiled Muslim women face anti-Muslim practices in everyday encounters but reveal a certain need to educate people against racism and Islamophobia as well as a certain lack of communication between the different communities in London. British people have learned to tolerate minority groups because it is politically and legally correct, but we may think that some have not really learned how to interact with them. It is as if British society lives in polite ignorance and indifference of the 'unfamiliar' other. In this sense, Hafsa (a 30-year-old British-Bangladeshi woman who wears the *hijab* or *niqab*) feels that racist and Islamophobic attitudes are more often the result of ignorance and misunderstanding than real rejection. She still believes in a positive outcome and explains the importance of non-Muslims interacting with Muslims:

> *Generally, people are open-minded and they are accepting once they get to know you, but… because they don't know you and they only see what's in the media, they have a perception of you which is hard for them to get rid of until they get to meet Muslims and interact with them. And once they do, most of the time it's positive, yeah. (Hafsa)*

Finally, veiled women avoid different spaces in Paris and London, but what is particularly interesting to note here is how the 'perceived' and 'real' geographies of Islamophobia match. More precisely, in Paris the geographies of Islamophobia perceived by the interviewees concern the central, the privileged and the crowded and unknown areas, corresponding to the quantitative data that show that the prestigious Parisian centre is the most affected part of Greater Paris. The map (in Figures 3.1 or 3.2) conforms to the qualitative mental maps shared by the Parisian interviewees (Najib, 2020a).

In London, veiled Muslim women avoid instead non-diverse areas (especially White areas), as well as streets with many bars and pubs and public transport. The connection of these perceived geographies of Islamophobia to the dark horizontal line on the map of London (in Figures 3.1 or 3.3) is less obvious, yet still presents similar elements that can be discussed here. First, even if there is Islamophobia in Muslim-majority areas, as shown in the quantitative data, it also occurs in predominantly White boroughs in the western part of Inner London. These areas are central and privileged, but they are not mentioned as such by the victims in discussing their avoidance; the reasons they give instead refer to race since they see these areas primarily as White.

Second, this western part is situated in the middle of the horizontal line and represents a transit zone that connects Muslim residential areas to the influential Inner London area. As this transit zone includes important transport networks and busy streets that veiled Muslim women tend to bypass and since these women cannot avoid their own residential areas, it can be said that the two geographies of Islamophobia in London ('perceived' and 'real') also correspond. Ultimately, French and British veiled women adopt specific and distinct spatial strategies, but in both cases they worry about their spatial mobility, and thus use more familiar spaces where they feel comfortable, safe and welcome. They feel particularly attached to their close residential area mainly because they meet the same people on a daily basis.

Spatial belonging and the significance of the home neighbourhood

People generally feel attached to their district of residence (Tajfel & Turner, 1986; Twigger-Ross & Uzzel, 1996; Manski, 2000), and this is particularly the case for victims of Islamophobia, especially veiled Muslim women (Kwan, 2008; Githens-Mazer & Lambert, 2010; Zempi & Chakraborti, 2015; Najib & Hopkins, 2019) who feel more comfortable and safer in their residential area. The sense of belonging to specific spaces gives an important meaning to feeling safe and 'at home' (Ignatieff, 1994; Hedetoft & Hjort, 2002; Isakjee, 2016). The way in which feelings of belonging are experienced by victims of Islamophobia depends on their situation of discrimination and reveals varying degrees of connection with different spaces.

By identifying the different spatial contexts to which the victim belongs (neighbourhood, district, city, region and country), the extent of their feeling was measured numerically at each spatial scale. More specifically, I asked the Parisian and London victims where they feel the most at home, and their answers were plotted on an ordinal numerical scale ranging from 1 (the least at home) to 6 (the most at home). The complexity and multiplicity of feelings of belonging reveal a struggle that can drive a victim to feel more attached to one spatial scale than another. Studies on the exclusion of Muslims in Paris and London have already shown contrasting patterns of attachment. For instance, Garbin (2011) explains how Muslims living in La Courneuve (a suburban municipality of Greater Paris) feel more attached to their town than to their region. Wacquant (2007) specifies that towns like La Courneuve represent relegated spaces occupied by relegated people who take root there (Wacquant, 2007). Likewise, in London, Datta (2009) highlights how the Muslim residents of Tower Hamlets develop strengthened ties with their borough, which represents a key focal point for their identity (because of their networks, friendships, etc.). These particular examples show how Muslims can feel that they are more Courneuvian than Parisian, or more residents of Tower Hamlets than Londoners. As for victims of Islamophobia, they have a stronger sense of belonging that is expressed at an even finer scale, as detailed in Table 3.1.

Strikingly, Table 3.1 shows little difference between the findings from the Parisian and London victims. First of all, both in Paris and in London, it can be seen that the finer the spatial scale, the greater the feeling of belonging. The average score is more positive in Paris than in London at each spatial scale apart from that for belonging to the country, for which the average score awarded by the Parisian victims is lower. Thus, the Parisian victims feel less connected to their country and more connected to their neighbourhood than the London victims. Do the French interviewees feel less French than the British interviewees feel British; and do the Parisian interviewees feel more anchored in their neighbourhood than the London interviewees? The corresponding differences (-0.12 and +0.21, respectively) are

Table 3.1 Extent of Parisian and London interviewees' spatial belonging

Average score	Neighbourhood	District	City	Region	Country
Parisians	5.17	5.06	4.73	4.16	3.24
Londoners	4.96	4.59	4.15	3.70	3.36

Note: The scores include all interviewees' answers, including those by men.

too small to be analysed in depth to compare the French and British cases. But I can say that the Muslim geographies in Paris – which may correspond to what we call the Parisian *banlieues* (Laurence & Vaïsse, 2007; Simon & Tiberj, 2013; Vieillard-Baron, 2004) with a specific social connotation synonymous with ethnic Othering, poverty and disadvantage (Cesari, 2005; Hargreaves, 1996; Dikeç, 2006) – certainly contribute to the development of stronger feelings of neighbourhood belonging.

In addition, it is true (to be discussed in more detail in the next chapter) that the French interviewees see the United Kingdom as being a more inclusive country for Muslims than France. From a general point of view, feeling French or feeling British refers to identity politics, and the interviewees mainly correspond to postcolonial identities who feel that they occupy a secondary place in France and in the United Kingdom. They suffer from significant stigma (related to insecurity, terrorism, etc.), which not only excludes them from Frenchness and Britishness but also from the influence of urban centres (especially those of Paris and London). They are pushed to the margins, towards their marginal and enclaved districts of residence, and even more towards the everyday geographies of their immediate home neighbourhood. Indeed, as they feel excluded from other outside spaces – including the transnational space related to their country of origin, which they do not necessarily see as a home space (Noble, 2005; Hage, 1998; Noble & Poynting, 2008) – they feel more included in their home neighbourhood, the only place where they feel themselves to be insiders.

In the examples of Paris and London, the district of residence rather refers to a borough or *arrondissement* which is a large area. But victims of Islamophobia feel safer in smaller areas such as their home neighbourhood. Because they opt to use spaces that they perceive as safe in order to avert potential situations of Islamophobia, this home neighbourhood becomes for them even more important in managing risk to personal safety and in creating safe spaces of inclusion and belonging. Therefore, the victims of Islamophobia in Paris and London feel spatially more rooted in their home neighbourhood, and some even avoid leaving it, as explained by Sabrina, Janna, Inès, Tasnim and Anila:

> Anyway, I stay not far from home, I go out no more than… not far from my home. So, these are the parks, there are some nearby, in my neighbourhood. In fact, I don't go there too much, I don't go outside too much. (Sabrina, a 42-year-old French-Algerian woman in hijab)
>
> Not necessarily. Frankly, apart from the nearby parks near my home for my children, yes I do a lot of activities with them. But no, I use, let's say, my little perimeter next to my home. (Janna, a 31-year-old French convert who wears the hijab or jilbab)
>
> There is another part which is more distant from my home where I can't even try [to visit]… when I go there alone, I can't, I always have to be accompanied in fact. (Inès, a 20-year-old French-Tunisian woman in hijab)
>
> I don't go out far in my neighbourhood. I mean, I was very freely open travelling and that until I had my kid [...] I don't actually go out by myself very

often anymore because of [Islamophobic] attacks. (Tasnim, a British-Indian woman in her twenties wearing the hijab or niqab)
    Yes, my neighbourhood is fine. There are no problems at all [...]. [But not in my borough] no I don't feel comfortable at all [...] I don't go far. It's just local run. So, I mean picking children up and from school and swimming and karate and... so that's it. (Anila, a 40-year-old British-Pakistani woman in hijab)

Here, the home neighbourhood refers to the streets, parks, and everyday places in which veiled Muslim women feel comfortable. This micro-locality is not free from Islamophobic incidents, but definitely corresponds to a safe space that fosters daily encounters and exchanges and prevents the proliferation of anti-Muslim acts, at least in the minds of these veiled women. People whom they meet in their neighbourhood are more likely to be people whom they know, such as neighbours or even friends. Thus, the home neighbourhood offers to these women an opportunity to find a sense of belonging, attachment, support and friendship. It shapes their religious identity more securely, and therefore means a symbolic spatial anchorage where they can live, experience and practise their Muslimness more openly. All told, we may wonder whether veiled Muslim women are familiar with any other spatial scale. Indeed, Islamophobia in a certain way encourages spatial inertia to the point where the home neighbourhood appears as a protective device for veiled Muslim women and an entire universe in itself. In the end, the home neighbourhood represents a sort of mental enclave – due to a certain state of mind (Wirth, 1998) shared by populations under the same threat – difficult to leave not just physically but mentally.

Conclusion

The social issues imposed by a multitude of segregation, deprivation, exclusion and insecurity systems are closely intertwined in urban and infra-urban Islamophobia. The political solutions proposed to combat these problems generally revolve around the principle of social cohesion (ensuring 'living together' and 'social peace'), which remains difficult to apply in practice, at the local scale. The mapping of Islamophobia in cities and districts (quite rare in Islamophobia Studies) has found socio-spatial effects that provide a critical vantage point from which to contextualize the sense of space and place. The spatial distribution of Islamophobia in the city is not coincidental and has emphasized many processes of oppositions, contradictions, associations, differences, similarities, rejections, repulsions and attractions.

    Using the examples of Paris and London, this work finds that the geographies of Islamophobia show a direct relationship with various spaces (geographical, socioeconomic, ethnic-religious) and can contrast with other studies, notably those carried out in the United States. First, geographical space is taken into consideration: in Paris (and to a lesser extent in London), Islamophobia occurs mostly in the centre, unlike other spatial contexts such as Chicago where the greatest occurrence of property attacks is in suburban areas (Cainkar, 2005), or in

San Francisco where Islamophobia is more apparent in suburban and even rural areas (Itaoui, 2020). Second, socioeconomic and ethnic-religious spaces are studied: in London (and to a lesser degree in Paris), Islamophobia is also important in deprived Muslim areas. Here, geographies of Muslim residency connect with spatial patterns of poverty, in contrast to the situation of American Muslims who are more likely to be part of the upper-middle class and rarely relate to the inhabitants of poor districts (Pew Research Centre, 2007; Sirin & Fahy, 2009; Marzouki, 2017). Important caveats relating to these conclusions with regard to Paris and London are discussed in this chapter, and show that the nature of an area as well as the specific places that it contains and the forms of mobility that it generates have a direct and important impact on Islamophobia.

Finally, urban and infra-urban Islamophobia brings to the fore the need for Muslims to leave their marginal residential areas for other spaces. Indeed, we cannot live in a world where various people must not cross each other's paths. Islamophobia affects ordinary people on a daily basis and particularly worries veiled Muslim women, who see their mobility being limited and their feeling of belonging damaged. In France, we know that Islamophobia is mainly an institutional issue, but in Paris and mainly Central Paris (where Islamophobia also occurs in public areas and on public transport), to this institutionalized discrimination is added a more physical Islamophobia linked to people's appearance. In the United Kingdom, Islamophobia seems to be essentially a public-attitude issue and involves zones of contact and encounter, such as the busy streets of London.

In response to these limitations, the Parisian and London victims have developed and adopted specific spatial strategies and belonging by avoiding unsafe spaces, streets and places. Their fear and anxiety reflect their perception of Islamophobia across the city and push them to the margins of society. These marginal areas are certainly considered by other identities as to be avoided, yet avoiding such marginal zones is much less restrictive than avoiding prestigious and busy central spaces. The trauma of veiled Muslim women resonates with their experience, and the never-ending potential threat of violence in public space fuels a sense of fear (Listerborn, 2015; Mansson McGinty, 2020) that can be felt at the scales of body and mind. This embodied and emotional Islamophobia leads Muslim bodies to reinvent new behaviours and practices. The following chapter will describe this in detail, without neglecting the diversity of all Muslim bodies, including veiled Muslim women and non-veiled Muslim women as well as Muslim men, who may also be deeply emotionally affected.

Notes

1 The French data are from the CCIF (which has a rich database at the individual scale as it takes all incident statements into account after verifying their content, unlike the Ministry of Interior's data, which are difficult to access and include only those that conclude in a formal complaints) and the British data are from the Metropolitan Police.
2 Greater Paris includes: 1) the city of Paris (with its 20 boroughs or *arrondissements*) known as *Paris intra-muros* or Central Paris; and 2) the inner suburbs (known as *proche banlieue*) comprising the three *départements* of Hauts-de-Seine, Seine-Saint-Denis and

Val-de-Marne. Greater London includes: 1) Inner London (with 14 boroughs, comprising the City of London); and 2) Outer London (with 19 boroughs). See the Appendix for full details.

3 While the term *banlieues* is used in common speech to refer to areas with a high concentration of poverty, social housing, minority groups, etc. (Cesari, 2005), here the inner suburbs (*proche banlieue*) refer to the three *départements* surrounding *Paris intramuros*. Therefore, this term 'suburbs' refers here to a range of areas (from the wealthiest to the poorest), not just degraded areas.

4 The variables are from INSEE, the National Institute of Economic Studies (Institut National des Etudes Economiques) which conducts the population census in France, as well as from DGFIP, the General Direction of Public Finance (Direction Générale des Finances Publiques), which manages professional, personal and patrimonial taxation.

5 The purpose of a typological map is to synthesize a large amount of information. Specifically, HCA is a factor analysis technique to group areas on the basis of similarity criteria and to construct a hierarchy. Here, the dendrogram highlights socio-spatial inequalities using five distinct classes.

6 In France, these areas are mainly known as ZUS (*Zones Urbaines Sensibles*) or 'troubled urban areas'. They were identified by French urban policy in the early 1980s on the basis of an aggregation of poverty factors (unemployment, social housing, low income, foreigners, educational backwardness, etc.). This classification is still relevant, even if it was superseded in 2014 by the priority zones identified by a single criterion: low income. The ZUS classification is still favoured in this research, since it better targets the territories associated with people with a postcolonial heritage (Desponds & Bergel, 2017) who are subject to significant discrimination, in particular Islamophobia.

7 Regarding the interviews in French, I translated into English only selected quotations for this book.

8 The *jilbab* refers to a long and modest garment worn by some Muslim women to cover the entire body apart from the face, the hands and the feet. This garment includes a headscarf.

4 Embodied and emotional Islamophobia

Introduction

As demonstrated, Islamophobia is expressed locally in everyday encounters, but it is first felt and experienced in people's bodies and minds. Indeed, the 'local' is infused with the 'personal', and our relationships with others are always affected by mood (Merleau-Ponty, 1962). Muslim bodies actively manage their impressions and feelings when engaging in public spaces where they feel less comfortable compared to private spaces. Through a qualitative people-based approach and literature on the geographies of embodiment, emotions, intimacy and Islamophobia (Ahmed, 2004; Valentine, 2008; Hyndman, 2004; Botterill et al., 2019; Mansson McGinty, 2020; Listerborn, 2015; Najib & Teeple Hopkins, 2020), this chapter reveals the behavioural strategies of French and British Muslim bodies and minds, as well as the existence of a more intimate Islamophobia (Moosavi, 2011).

First, Islamophobic incidents are exacerbated by physical markers of Muslimness that generate negative reactions (Cainkar, 2005; Allen, 2010; Najib & Teeple Hopkins, 2020). In response to these anti-Muslim reactions, Muslim bodies (especially the more visible ones) make several changes to their everyday behaviour. They constantly hide or adjust their religious identity by acting or dressing differently in various contexts (Siraj, 2011, Dwyer, 2008; Sirin & Fine, 2007; Najib & Hopkins, 2019). Ways of speaking, dressing, eating and (re)presenting oneself are obviously personal practices, but they are governed by an Islamophobic climate and a political and social system that dictates whose the 'normal' bodies are and how they should behave (Ahmed, 2000). In order to reduce anti-Muslim hostilities, highly stigmatized Muslim bodies often perform invisibility and normalcy (Hopkins, 2007; Siraj, 2011, Dwyer, 2008; Carr, 2016; Najib & Hopkins, 2019; Khan & Mythen, 2019). They reinvent new embodied practices and pursue these behavioural strategies when negotiating public Islamophobia.

Also, Muslim bodies and minds develop significant internal fears and anxieties that impact their sense of space. Their sense of security is influenced by Islamophobic violence (soft or not) that affects the public expression of their Muslim identity, sometimes pushing it into the home space. Therefore, their home becomes an affective space in which Muslim bodies feel safe and perform intimacy by avoiding leaving it as often as possible. But this enclosed intimate space is not

DOI: 10.4324/9781003019428-4

free from Islamophobic incidents. Indeed, Islamophobia can also take place at home and thus affect kinship relations (Iner & Nebhan, 2019; Mansson McGinty, 2020). Family members who have dissimilar religions or interpretations of Islam may develop Islamophobic attitudes towards Muslim relatives who often try to avoid significant disputes (Ramahi, 2020; Abbas, 2019; Bolognani, 2007). Finally, these strategies show the sophisticated responses of Muslim bodies in the growing context of racism, terrorist threat, politicization of the Islamic faith and an uncertain future (economic, health, social, etc.).

Spatialized Islamophobia embedded in the body and mind

Our relations with others are always 'mooded', whether positively (enjoyment, desire, etc.) or negatively (anxiety, fear, etc.) (Merleau-Ponty, 1962). Moods and emotions are fundamental human attributes and, in the case of Spatialized Islamophobia, everyday encounters are rather associated with the notion of 'feared others' (Ahmed, 2000). The emotional environment of mainstream society determines what fear is: it is an unpleasant emotion caused by anticipation or awareness of risk and danger. Fear intersects between the personal and the societal, the emotional and the rational (Pain & Smith, 2008), and is therefore written on the body. Emotions and embodied experiences of fear, particularly of Islamophobia, are shaped by processes of Othering (Ahmed, 2000; Mansson McGinty, 2020; Pain, 2009) mainly spread as globalized processes. The construction of Otherness is communicated and mediated through global media discourse, and it is generally used to legitimize discrimination and racism. Indeed, Islamophobia is globally portrayed in interaction with global geopolitical events, and this global Islamophobic industry is directly embedded into bodies and minds (Mansson McGinty, 2020; Botterill et al., 2019; Listerborn, 2015; Najib & Hopkins, 2019). The connection of global Islamophobia to emotional Islamophobia brings its spatiality to life and deserves to be explored.

Geopolitics has a significant responsibility for everyday experiences, and critical social sciences (in particular critical geography) must take greater account of the everyday, the embodied and the emotional, as advocated by feminist academics (Hyndman, 2004; Sharp, 2007; Ahmed, 2004; Rose, 1993; Pain & Staeheli, 2014; Pain et al., 2010). By constantly juggling the 'private' and the 'public', they have been able to show how the everyday experiences of bodies and minds relate to geopolitical discourses and events. Consequently, this chapter takes everyday life, body and emotions deeply seriously.

First, Mansson McGinty (2020) explains that the study of the everyday realm when examining any geopolitical phenomenon (such as Spatialized Islamophobia) is essential. She shows, through feminist geopolitical theories, that such racism is lived, felt and embodied in people's everyday lives. It can affect their self-confidence and their sense of belonging to society. Second, the body, according to Merleau-Ponty (1962), is part of a social realm based on perception, practice and bodily moves. Bodies can be recognized as being similar or different, familiar or strange, and can thus describe a sense of inclusion or exclusion. A body can feel either at ease or out of place because it functions as a marker of social, cultural and religious differentiation. Third, Ahmed (2004) shows that emotions are not simply feelings between individuals or groups: they are also instruments of governance that are constituted by and constitutive of social relations. Emotions are also fluid and fleeting (Harrison, 2000), since they move between bodies and do not reside in the same individual in fixed forms.

The study of embodied and emotional Islamophobia implies a better approach to the relationship between spaces and identities. Spatialized Islamophobia links considerations of spatial patterns to theories of intersecting identities (Najib & Hopkins, 2020). This anti-Muslim racism can easily be read in nationalist political discourses, and it is also lived and experienced in the daily life of the most vulnerable bodies. In most spatial contexts, the main people affected by Islamophobia are veiled Muslim women, who became the epitome of embodied geopolitics (Dixon & Marston, 2011; Hyndman, 2004; Williams & Massaro, 2013) mainly because of their great visibility. Indeed, being a visible Muslim woman is a geopolitical issue (Najib & Hopkins, 2019). Mainstream media and politics report an increasing visibility of the *hijab* in the public sphere, more than any other symbol of piety or personal belief and identity. As a result, the *hijab* has become a strong marker of Muslim identity, and its visibility in public spaces a strong synonym for Spatialized Islamophobia.

In addition, there is another concern, specifically gender, related to the feminist struggle for women's right to control their body (hooks, 2000). In this male-dominated society, veiled Muslim women appear as marginalized groups who need to be carefully managed (Bilge, 2010; Farris, 2017; Dwyer, 1999; Kwan, 2008; Fernando, 2009; Mirza, 2013; Hopkins, 2016), and feminist scholars in Sociology, Criminology and Geography have played a particularly prominent role in examining the interactions between religion, gender and place and in shaping how social politics produce emotional landscapes for marginalized groups (Pain, 2001; McDowell, 2008; Valentine, 1989). Veiled Muslim women are seen as universally vulnerable, and the study of their everyday lives is naturally a pressing issue. Exploring feminist geopolitics through the lens of bodies and minds amounts to taking an interest in their everyday fears and concerns in response to Islamophobia. It is therefore important to analyse in this chapter the embodied and emotional geographies of veiled Muslim women's experiences of Islamophobia, as well as their ways of negotiating their Muslimness.

Having said that, Muslim men, too, worry about Islamophobia. Even if they do not necessarily acknowledge being afraid for themselves, they express fear for their family and especially for the women of their family (daughters, nieces, wives, sisters, and mothers) (Najib & Hopkins, 2019). Muslim men particularly suffer from significant demonizing stereotypes related to violence, patriarchy, sexism and especially terrorism (Staeheli & Nagel, 2008; Hopkins, 2007; Cohen & Tufail, 2017; Mythen et al., 2009). Ahmed (2004) and Swanton (2010) have explained this idea of affective circulation between bodies, and how – in the 'war on terror' era – the racist judgements of 'becoming terrorists' take place particularly in moments of encounter. Muslim men are subject to important scrutiny and feel fearful of being scapegoated and suspected of potential terrorist activities. Negative stereotypical views of Muslim men (Archer, 2009; Runnymede Trust, 2017) also represent an important political concern that needs to be combatted. More importantly, it is thus necessary to remember that Islamophobia has a marked psychological influence on all Muslim populations, including men (Mac an Ghaill & Haywood, 2015; Hopkins, 2007), children and teenagers (Elkassem et al., 2018; Younus & Mian, 2018; Sirin & Fine, 2007) and both veiled (Gökariksel & Secor, 2012; Najib & Hopkins, 2019; Listerborn, 2015; Bilge, 2010; Dwyer, 1999) and unveiled Muslim women (Mir, 2014; Perry, 2014; Kwan, 2008; Selod, 2015).

All told, the public's ignorance of Islam and the political prominence of the 'Muslim problem' (Dunn & Kamp, 2009) directly affect Muslim bodies. Rather than experiencing fear themselves, people view Muslims as a source of fear (although see Day, 1999; Hopkins & Smith, 2008). This demonizing of Muslim bodies in the media and politics, in turn, affects Muslims' own sense of security. In previous chapters, I have explained that Muslims are neither failed citizens (Chapter 2) nor self-segregated residents (Chapter 3), challenging the negative mainstream representations. This chapter is about demonstrating that Muslim bodies (veiled bodies, full-covered bodies, bearded bodies, etc.) are not dangerous and, on the contrary, feel themselves insecure in many spaces. Depending on bodily markers, some Muslims may be feared to a greater or a lesser extent in various places and times; which makes them more fearful. In this sense, Muslim bodies become at the same time subjects of hatred and of fear (Kwan, 2008), and various power relations attempt to differentiate between 'the feared' and 'the fearful'.

Indeed, there are important power dynamics that depict the Islamic faith negatively and therefore push its followers to seek more security. The emotional trauma resulting from Islamophobic discourse has been reiterated so often that Muslim bodies end up feeling constrained by their religious identity. Subsequently, Muslims have no other choice but to negotiate their religious identity at a personal level through specific behaviours and practices. Even though religion represents a major part of their identity and is an important driver in how they live their lives (Peach, 2006a; Modood et al., 1997), Muslim bodies still struggle to control their fears, hopes, routines, domestic environments, and so on. In the end, this global fear of Islam seriously challenges the way in which they behave, interact with and engage in public and private spaces.

From public to private spaces: Muslim bodies performing invisibility, normalcy and intimacy

Behaviours and spaces are mutually dependent (Ardener, 1981; Rose, 1993). Indeed, space has a symbolic meaning, and its complex impact upon our behaviours is mediated by specific cognitive processes (Najib, 2018). To understand them better, it is important to consider human bodies and minds as well as their functions in our environment and everyday life. Bodies and minds can be animated by feelings of discomfort and insecurity that affect our experiences of public and private spaces. Here, the connection between fear and the spatial environment requires the use of and a better appreciation of embodied and emotional geographies. Emotions are at the heart of how we perceive the immediate environment, and the body diffuses them in the form of behaviours and attitudes (Damasio et al., 2000; Milton & Svasek, 2005; Paterson et al., 2019).

As demonstrated throughout this book, Islamophobia is undeniably spatially located, and its manifestation in the space provokes a moral panic that leads Muslim populations to negotiate, through 'performative bodies' (Butler, 2011b), their own everyday geographies; that is, to develop a safer environment by performing strategies of invisibility, normalcy and intimacy. They develop specific strategies to deal with the securitization of their bodies when occupying various spaces. Negotiating between and across bodily and spatial difference reveals how Spatialized Islamophobia is encoded with meanings of space and presence in everyday interactions. The concerns of Muslim bodies over their security and potential exposure to violence push them to consider carefully their spatial environment and their interactions with people in public space.

The ways in which Muslim bodies behave in public spaces refer to the notion of risk (whether real or imagined), closely bound up with spaces, geopolitical events and human anxiety. Risk mainly concerns women, minority groups and youth, and how they conduct their everyday lives, their interactions with others and their relationships to their bodies. (Rose, 1993; Brah & Phoenix, 2004; Pain, 2001; Valentine et al., 1998; Scott, 2003). There are scientific works on the experiences of Muslim bodies and the way in which they manage aspects of risk as part of their everyday spaces (Dunn and Hopkins, 2016; Kwan, 2008; Najib & Hopkins, 2019; Itaoui, 2016; Mansson McGinty, 2014), revealing specific behaviours and practices that vary according to personal characteristics such as gender, social position and level of education, but also personality and clothing. For example, we know that public space is primarily constructed as masculine (Rose, 1993; Bondi & Rose, 2003), and even veiled Muslim women are aware that some of their practices are foremost due to their gender and secondarily to their headscarf. Veiled Muslim women's fears and attitudes are well documented (Bilge, 2010; Dwyer, 1999; Kwan, 2008; Fernando, 2009; Mirza, 2013; Gökariksel & Secor, 2012; Najib & Hopkins, 2019; Listerborn, 2015), but general studies on behavioural changes in public space in line with risk, safety and danger focus on all women (Rose, 1993; Valentine, 1989; Pain, 1991, 2001; Farris, 2017).

The literature on feminist geography has also explored the impact of women's fear on their behaviours, and mainly refers to their sense of space as difficult (ibid.). Indeed, space can be unpleasant and oppressive for women who can feel confined and constrained primarily by their gender. In this sense, veiled Muslim women undoubtedly feel a double constraint that they have to manage in public spaces. For example, they have to monitor and control their own bodies and behaviours in public spaces, just like other women especially with regard to time-geography (Hägerstrand, 1970; Rose, 1993) and everything else associated with women. Additionally, they have to manage an Islamophobic threat because of their headscarf. To be more exact, when veiled Muslim women avoid going out at night alone, it is mainly due to their gender since most women do the same (Rose, 1993; Bondi & Rose, 2003). But when veiled women avoid going outside for several days immediately after a terrorist attack, this is clearly related to the great visibility of their religious identity and, above all, their *hijabs* (Najib & Hopkins, 2019). Like everyone else, these women learned at an early age (especially from their parents) where the risks lie, and those risks partly changed when they decided to wear the headscarf. Indeed, these women are not born with a *hijab* around their heads, and they notice a clear change upon starting to wear one (ibid.). Before wearing the *hijab*, the fear is more a matter of physical and sexual assault (and even racist attack), but the veil adds a new risk. The global fear of terrorist attacks and the rise of political Islam has created a suspicion that is often reflected in public opinion against *hijabi* women, leaving them particularly vulnerable. That said, Muslim women who wear no headscarf also have to deal with fears and concerns related to Islamophobia, which are just as real and serious. Therefore, Islamophobic hostilities affect all Muslim bodies and push them to express their Muslimness less overtly.

First, Muslim bodies can pursue strategies of invisibility when negotiating their Muslimness. The ways in which they use public spaces compel them to be less open and visible. More precisely, they change or conceal their religious identity in various contexts to avert or lessen potential situations of Islamophobic discrimination (Hopkins, 2007; Siraj, 2011, Dwyer, 2008; Carr, 2016; Najib & Hopkins, 2019). They can erase markers of Muslimness or change their everyday routines and behaviours, especially those who are less confident due to anxiety, paranoia and depression (Samari, 2016). These will prefer to be less visible by avoiding bold behaviours that can provoke strong reactions.

In addition, this desire to be less visible may be motivated simply for security reasons. Some Muslim bodies are well aware of the hostility that they arouse, and therefore, prefer to avoid putting themselves in dangerous situations even if they do not have trust issues. For example, they hide their *hijabs* under their hooded coat; they wear a simple turban instead of a *hijab*; they rely more on their car and avoid public transport; they do not partake in any activity alone; they sometimes wear trainers in case they have to run; they read information on Islamophobia in order to be better prepared; they limit and time their activities; they say that they are vegetarian instead of saying that they eat only *halal* meat; they avoid reacting to racism for fear of escalation; they over-apologize to avoid confrontation; they avoid eye contact; they wear their Islamic clothes only inside the mosque; they avoid speaking about religion; they dress, behave and speak as non-Muslims; they conceal everything that refers to Islam, and so on (Sirin & Fine, 2007; Najib & Hopkins, 2019; Hopkins & Clayton, 2020).

Muslim bodies may identify specific risks and dangers associated with Islamophobic sentiment and find a way to manage them through constant behavioural negotiations, which are primarily rooted in power relations. Indeed, as previously explained, Spatialized Islamophobia is linked to power relations that force Muslim bodies to remain discreet if they feel unconfident or badly judged, or even in danger; and conversely to feel more comfortable when interactions take place between people sharing a common identity and opinions.

Second, Muslim bodies can pursue strategies of normalcy when negotiating Islamophobia (Khan & Mythen, 2019; Mythen et al., 2009). Following the moral panic caused by terrorist threats and the politicization of their faith, Muslim bodies are seen as strange (Allen, 2010; Bonn, 2012), in the same way as those associated with other social problems (such as gangs, drugs, prostitution, etc.). Consequently, while expressing their religious identity, Muslim bodies want to show that they are normal by conforming as far as possible to the hegemonic norms and expectations (Macdonald, 2006). In this context, public space becomes a space through which the fear of strange bodies (Ahmed, 2004) and their potentialities can justify their marginalization and exclusion. The circulation of vulnerabilities and racisms that can be found during everyday encounters (Wilson & Darling, 2016) pushes Muslim bodies to adjust themselves as much as possible. Indeed, these vulnerabilities and racisms influence and affect their behavioural expression by accentuating their ordinariness.

To challenge and reduce the Othering gaze (which describes Muslims as strangers, foreigners, inferiors, dangerous), they choose to perform normalcy (Khan & Mythen, 2019) by speaking 'normal', looking 'normal', dressing 'normal', thinking 'normal', consuming 'normal', eating 'normal', reading 'normal' and travelling 'normal'. To undermine popular perceptions, they show that they, too, speak perfectly the national language of the country in which they live; they also wear Western clothes; they read Western philosophers; they eat local products, etc. They want to show that they are 'normal' citizens like any other, and, even more so, they want to show that they are good integrated citizens (Fernando, 2009; Ahmed, 1992; Gökariksel & Mitchell, 2005; Bowen, 2007; Dwyer, 2008).

This expression of normalization, goodness and integration does not equate to assimilation (Dunn & Kamp, 2009), since Muslim bodies want to demonstrate that they are ideal citizens (courteous, well-mannered, pleasant, polite, loyal, upright, hard-working, ambitious) at the same time as being good Muslims. Muslim bodies want to challenge the representation whereby anything that deviates from what mainstream society considers 'normal' is necessarily dangerous, by going far beyond normalcy and, in fact, by performing excellency. They usually try to manage multicultural intimacies through engagement with others by smiling at them, greeting them, helping them, and so on (Najib & Hopkins, 2019). Indeed, Muslims become 'ambassadors of Islam' (Khattab & Modood, 2015; Hussain & Yilmaz, 2012), adopting a proud attitude and presenting Islam deeply positively. Despite their great efforts to appear 'normal' and 'ideal' citizens, Muslim bodies often continue to feel out of place and unaccepted in many public spaces. Their feelings of marginalization and exclusion from the public space and opinion are still present, and they may even feel pushed into more hidden and private spaces, such as the home space for example.

Third, we understand here that, for victims, public and private spaces have different meanings, since Islamophobia is deemed to be almost 'normal' in public spaces and part of everyday life. Most Muslim bodies are aware of this normalization and decide to pursue strategies of intimacy in response to this public Islamophobia. They prefer or, to be more exact, feel compelled to use more private spaces, be it their own home, the home of their friends/neighbours/relatives or their own car, and so on (Kwan, 2008; Bayoumi, 2010). Remember Chapter 3 (urban and infra-urban Islamophobia), where I explained that the majority of Muslims (primarily veiled Muslim women) feel safer in their home neighbourhoods. They consider the area immediately surrounding their place of residence (house or apartment) to be safer. But even the streets of this small zone can be a source of anxiety for some Muslim residents, who are still likely to experience Islamophobic racism, harassment and discrimination. Therefore, for Muslims who seek intimacy, tranquillity and security, the home is the ultimate secure space (Kwan, 2008). Indeed, when people feel excluded from many public spaces and places, their home is of great importance (Noble, 2005).

Social fear and anxiety about going outside can increase quickly, especially among Muslims who have experienced a traumatic Islamophobic incident such as physical aggression, or following a global terrorist attack (Mansson McGinty, 2020; Najib & Hopkins, 2019). Their psychosocial state can lead them to cut themselves off from others by staying at home most of the time. This may well be the case for Muslims who claim to follow an orthodox Islam, since for religious reasons they first differentiate themselves from contemporary society and avoid involvement in civic or political life. Because of their highly visible Muslim clothes and practices, they may regularly suffer physical or verbal Islamophobic attacks and, for these reasons, may prefer to isolate themselves at home.

Generally speaking, the home space is very important to Muslims (Noble, 2005; Hopkins, 2007), especially for Muslim women (Dwyer, 2000; Mohammad, 2005; Phillips D., 2009). They often identify with the private space of their home, and some are convinced that it is the only space where they can self-realize and emancipate themselves. Here, a clear contradiction appears between the public space where these women feel more vulnerable and the private space where they are comfortable. Certain secular laws (i.e. anti-headscarf and anti-*niqab* laws in public spaces in many European countries) have pushed them into more spatial isolation (and even social, educational, professional and economic isolation). These laws have widened the divide between public and private spaces by allowing only in private the wearing of symbols of Islamic belonging (Teeple Hopkins, 2015), without really appreciating that veiled women rarely wear their veils at home where they are not in contact with men outside of their immediate family. Instead of fighting against their social and spatial exclusion, these national laws actually worsen the living conditions of veiled Muslim women and contradict the arguments about their supposed self-segregation (Phillips, 2006; Mondon, 2015; Dunn et al., 2015; Simon & Tiberj, 2013). Geopolitical events also affect their relationships with public and private spaces, and their emotions can push them to see their home as a traditional refuge in times of distress (Dwyer, 2000).

All things considered, the home is thought of as an everyday space of symbolic significance where one can develop a sense of self, identity, belonging, empowerment, community and connectedness (Rose, 1993; Blunt, 2005; Warrington,

2001; hooks, 1990; Gregson & Lowe, 1995; Dwyer, 2000; Phillips D., 2009). It is, especially for Muslim minorities, the space where religious, cultural and life values are transmitted between the generations as well as where they can see themselves as agents of their choice and destiny, compared to an unflattering status as an oppressed religious minority group. But the problem is that the home can also be a space where Islamophobia manifests itself and where some family members use their influence to dictate their anti-Muslim views.

Islamophobia in the home space

Islamophobia may occur in the private space of the home and be expressed directly within the family and by Muslims themselves. In general, debates about Islamophobia focus mainly on the global and national scales, as well as on the streets and other public places. But whether in a national emergency or at home, Islamophobic threats are just as serious. Indeed, private spaces such as the home, rarely discussed within Islamophobia Studies, are also of crucial importance (Mansson McGinty, 2020). Islamophobia within the family (close or extended) is beginning to receive more attention, and recent studies analyse inter-family criticisms and conflicting kinship relations between religious Muslims and certain family members, whether they are non-Muslim (Mansson McGinty, 2020; Ramahi, 2020), secular Muslim (Yucel, 2010) or simply Muslim with a differing level of religious practice and interpretation (Iner & Nebhan, 2019; Abbas, 2019). Family ties can foster Islamophobic incidents in the intimate space of the home due to ignorance of Muslim practices or dissimilar points of view. The domestic space is not only an emotional space with a primary role in shaping social and family relations (Rose, 1993; Blunt, 2005) and Muslim identity (Dwyer, 2000; Mohammad, 2005; Phillips D., 2009); it is also a driver of everyday Islamophobia, which is probably more difficult to overcome than other forms since it involves family members.

One of the most recurrent Islamophobic remarks at home concerns the *hijab* worn by young girls, mainly disapproved of by their parents or other relatives (Moosavi, 2011; Mansson McGinty, 2020; Yucel, 2010, Iner & Nebhan, 2019; Abbas, 2019; Shterin & Spalek, 2011). Here, family members who have differing religions or interpretations of Islam may develop strong hostilities to the *hijab*. Their anti-*hijab* sentiments are influenced by mainstream media and politics, and are expressed within interfaith or even Muslim families. Family members can blame relatives for wearing the headscarf using well-known stereotypical arguments relating to oppression, sexism, passivity and radicalization (Staeheli & Nagel, 2008; Farris, 2017; Frankenberg, 1993; Dwyer, 1999; Najib & Hopkins, 2019). Others are reluctant to see their young relatives (mainly daughters and nieces) with a *hijab* in the belief that they are too young and immature to truly understand its religious meaning. Some may dislike the style of head covering that these young women choose to wear, preferring a simple turban instead of their *hijab* or *jilbab*. There are also Muslim families who may disapprove of the headscarf mainly for security reasons, because they are afraid of *hijab*-motivated crimes. For all these reasons, the *hijab* can cause significant conflicts in family relationships.

Further important Islamic markers are also disputed by family members, and these are even more visible in the private space of the home than in public spaces. For example, prayer: Muslims pray five times a day, and trying to hide this practice from their families is in most cases impossible. Prayers can be seen as radical Islamic practice by some people (non-Muslims, secular Muslims, etc.) and, as a result, some practising Muslims living in a hostile family environment have to deal with negative remarks and feelings on a daily basis. Simply because of their worship practice they may be scrutinized or considered suspect by their own family, even if they too are Muslim. These conflicts arise because parents worry about any form of radicalization and wish to make sure that their children understand their religion properly (Abbas, 2019). Thus, Muslim youth (men and even women) can be suspected of potential terrorist activities by their own family (Hamid, 2011; Saeed, 2016; Pearson & Winterbotham, 2017; Saeed & Johnson, 2016). Their relationship to religion is therefore questioned as their families do not want them to become too religious or do not respect their religious identity (Abbas, 2019; Mansson McGinty, 2020).

Some converts may even suffer from a double Islamophobic-type hostility: the first from their non-Muslim family who do not tolerate their conversion or think that it is only a fad or a phase; and the second from their Muslim entourage or in-laws, who underestimate their knowledge of Islam and make fun of their religious practice and their level of devotion (Moosavi, 2011). This double rejection can lead them to become 'a minority within a minority', as noted by Roy (2013: 178). Besides, there are other problems, mainly involving extended family and specific to the *halal* diet that Muslims follow (no pork, only *halal* meat, no alcohol, etc.). Some Muslims are uncomfortable with certain intolerant family members and avoid visiting them, especially for family meals. They are particularly afraid that their children will eat prohibited foods, and therefore avoid stays with them through a lack of confidence.

All these examples are real (Mansson McGinty, 2020; Yucel, 2010; Iner & Nebhan, 2019; Abbas, 2019; Ramahi, 2020), and here Muslims are confronted with what Moosavi (2011) in his work on Muslim converts calls the 'intimate Islamophobia'. I refer to this notion by extending its scope not only to Muslim converts but to all Muslims living in a hostile family environment. For these Muslims, there is probably no difference between what they may experience in public spaces and what they live through in their private homes. Here, the boundaries between public and private spaces become blurred, while mainstream studies expose the experiences of Muslim bodies suffering from discrimination essentially only in the public sphere. Yet, Islamophobia within the family should not be overlooked because family, as a model of government (Foucault, 1991), can cause major psychological trauma that takes a lifetime to overcome completely.

Ultimately, what is so striking here is how, notwithstanding, media-influenced Islamophobic hostilities manage to penetrate the homes of Muslims and undermine family relationships. The connectedness between global, national, urban, infra-urban, embodied and emotional Islamophobias is spatially readable and shows how Spatialized Islamophobia can also reach the intimate space of the home from an 'inside' environment (i.e. directly from Muslims themselves and the family), leaving its victims without any respite. The study of embodied and emotional Islamophobia has allowed scholars to reveal the presence of this specific type of Islamophobia that is experienced in the home space which, surprisingly, also becomes a space of resistance to anti-Muslim hatred. And more definitely, the study of embodied and emotional Islamophobia has allowed me, by listening to the testimonies of the Parisian and London victims, to better understand how Islamophobia can affect their sense of self and belonging on a daily basis. Anti-Muslim hostility makes them deeply apprehensive about expressing their Muslimness in the public sphere, and the sociality of everyday life forces them to create, develop, adopt and negotiate new behavioural changes.

Behavioural changes for Parisian and London Muslims

Left to take care of their experiences of and exposure to anti-Muslim racism, the Parisian and London victims of Islamophobia react by altering their everyday

Figure 4.1 Examples of clothing for Muslim women
Note: Sarah is a French Muslim convert who wears different outfits according to her needs and desires. She can wear: 1) a chic abaya; 2) fashion clothes with a turban/hijab; 3) a jilbab; 4) a djellaba with a hijab; 5) a sari; 6) Western clothes with a turban; or 7) Western clothes with her hair visible.

behaviours and practices in order to reduce the potential threat. Individual interviews are rich sources of information, because oral histories are imbued with emotions, feelings and attitudes and are situated in a specific socio-spatial context. Indeed, person-centred interviews allowed me to collect the emotional and embodied reactions of the 60 participants (33 in Paris and 27 in London). The vast majority of the participants wear a visible marker of Muslimness, whether a beard, a veil or an Islamic garment, as detailed in Table 4.1. More exactly, of the male interviewees (6 in Paris and 4 in London), two in Paris and one in London wear no visible marker of Muslimness (besides skin colour or name), and similarly, of the female interviewees (27 in Paris and 23 in London), only one French and three British women wear no veil. The rest of the women interviewed, both in Paris and London, don headscarves and in great majority *hijabs*. In the following sections, I focus on all Muslim bodies (women or men, whether or not veiled or bearded) and how they negotiate important transformative behaviours either by hiding or regulating their religious identity in specific situations.

Table 4.1 The 60 interviewees' Muslim visibility

	Paris		London	
	Men	Women	Men	Women
Interviewees	6	27	4	23
Visible marker	4, comprising:	26, comprising:	3, comprising:	20, comprising:
	• 1 beard/ qamis*	• 1 turban/ bandana	• 1 beard/ thobe*	• 1 turban/ bandana
	• 3 beard	• 17 *hijab*	• 2 beard	• 8 *hijab*
		• 3 *jilbab*		• 1 *jilbab*
		• 3 *hijab*/ turban		• 4 *niqab*
		• 1 *hijab/jilbab*		• 1 *hijab*/ turban
		• 1 *jilbab*/ turban		• 3 *hijab*/ *jilbab*
				• 2 *hijab*/ *niqab*
No visible marker	2	1	1	3

Note: A *qamis* is a long garment worn by Muslim men, usually for praying at the mosque. This term is used by French interviewees, while British participants use the term *thobe* instead. Both terms correspond to the exact same garment.

Towards more invisibility

Thanks to the many interviews I conducted with victims of Islamophobia, I can indicate specific behavioural practices and strategies of invisibility. First of all, in response to Islamophobia Muslim bodies (men and women) try to be discreet. Some are not necessarily confident in specific contexts and are afraid of potential repercussions if they openly show that they are Muslim. This is particularly the case of Linda (a 31-year-old French-Algerian woman in *hijab* or turban) who, after experiencing an important instance of Islamophobic discrimination in the workplace, decided to act as a non-Muslim in a new job. In her new professional environment, Linda does not wear a headscarf and does not speak at all about her religion. She does not say that she is a Muslim because she is afraid of not being accepted and integrated into her workplace (even losing her job). She explains the following:

> *In my [former] work [...] I have been looked at askance, I felt like oppressed. It's like we have to be little mice actually! Now I have a new job, I've been there for 4 months. I don't speak anymore... it's really something I do now, I don't talk about my religion, I don't say that I go to the mosque, I don't say that I am veiled. Whereas, before, in my old job, I was naive, I spoke more openly, I used to say to my colleagues that I go to the mosque, I used to explain what Ramadan is... But now, there is a break, I never want anyone to know what I do in my private life, in everything that is related to religion, and all that because it has really damaged my former position. (Linda)*

Now, Linda erases everything that refers to her religion in her new workplace where she feels totally integrated, while in her private life she appears more religious because she dons the *hijab* and the *abaya*.[1] She is aware that this situation can only be temporary, since she knows that it is not necessarily good for her mental health.

Other Muslim bodies become more discreet simply for security. It is not necessarily because they are less confident, but rather because they are well aware of the danger in certain situations. For example, some veiled Muslim women told me that they have become more worried about the consequences of Islamophobia since they became mothers. This is the case of Zahra and Tasnim, two young mothers who now feel more vulnerable to physical assault, as in the following excerpts:

> *Yes... Yes, I'm afraid of being assaulted, yeah. It's especially since I'm a mom, since I have a stroller or a baby in my hands, I pay much more attention to the places I go, to the hours. (Zahra, a 30-year-old French-Moroccan woman in hijab)*
> *I mean, I was very freely open travelling and that until I had my kid and I got attacked. (Tasnim, a British-Indian woman in her twenties wearing the hijab or niqab)*

More precisely, in trying to be less visible to avoid Islamophobic situations, Muslim bodies often do so through their clothes. Indeed, they mainly adapt their clothing according to others' reactions and the spaces or places that they use. Although veiled Muslim women are most concerned about dressing strategies especially because of their headscarves, Muslim men also avoid wearing specific Islamic clothes such as the *qamis* or *thobe* (see note of Table 4.1). This is the case for Wassim and Aboumir, two French-Algerian men in their thirties who have short beards and avoid the *qamis*:

> *Even in my dressing, I dress a lot less like that [a white qamis]. Here, I came to see you like this because I wanted to come like this because… But, I'm usually more careful. (Wassim)*
>
> *Yeah before, I had a full beard, I went out quite frequently, 95% of the time, in a qamis. As you can see today, this is no longer the case. But there is also a religious posture which has changed, religious opinions which have changed, but there is also a part of being more comfortable like this [without a full beard and a qamis]. (Aboumir)*

In Britain, the Muslim men whom I interviewed do not avoid wearing Islamic clothes, but one noticed in his local mosque that the great majority of men who pray with him wear a *thobe* only inside the mosque. He details:

> *I know that some Muslim people, men, are worried about wearing Islamic clothing, when they go from home to the mosque for example. So, they wear it [the thobe] only inside the mosque. I know some people, they are worried that any attack, or anything like that… I know a woman… I've seen a lot of sisters, they wear the hijab, but they wear a hoodie on top, or hats. I have seen a lot of sisters doing that. Men, I haven't seen much, to be honest. There was a few. Not many. (Jamal, a European resident with Eritrean origin, in his thirties who has a small beard)*

At the end of his comment, Jamal explains that strategies about dress are more common in veiled Muslim women than in men. This reality, well documented in the social sciences, is that since veiled Muslim women are more visible and more vulnerable, they represent the perfect target for anti-Muslim motivations. They are well aware of this danger, and that is why they try as far as possible to cover or hide their headscarves. Indeed, for these women, wearing a *hijab* is challenging (Bowen, 2007; Scott, 2007; Fernando, 2009), whether at school (Dwyer, 1999), in the workplace (Lewis, 2009; Rootham, 2014), in the media and fashion industry (Warren, 2019; Kilicbay & Binark, 2002) or in public spaces (Secor, 2002; Listerborn, 2015; Gökariksel & Secor, 2012).

Of the veiled participants, Parisian women have developed more strategies than the British participants. As detailed in Najib and Hopkins (2019), veiled Muslim women adapt their clothing according to where in the Paris region they need to go and the people whom they are likely to meet. They try to be less visible, especially in professional and institutional contexts. To make a good first impression, they will tend to wear a simple turban when they complete administrative paper-work in public institutions or go for job interviews, as explained by Habiba and Emilie:

> *It depends, in fact it's random. But [...] when I have an administrative appointment, I will wear it [the turban] yes, yes, yes... In fact, in all adminis-trative places since I would be too afraid to come across a racist person who has prejudices and who blocks my [administrative] procedures. So, I put on the turban systematically. (Habiba, a 29-year-old French-Moroccan woman who wears the hijab)*
>
> *If they see my neck [during a job interviews], maybe they'll see my brain a little bit more! So, these are strategies [...] You have to show that: 'Hey, I'm Muslim, but I'm fine, I live well, I'm open, everything is fine, I'm going to do my job'. It's true that we have to redouble our efforts to leave a good impression and to show that we're fine, we have two arms, two legs. (Emilie, a 29-year-old French convert in hijab)*

In their workplaces, it is even clearer; almost all opt to wear a simple turban with Western clothes (Najib & Hopkins, 2019). In this sense, Camélia and Fatou explain that they feel that they are transforming themselves, one referring to dis-guise and the other to a chameleon:

> *Today when I go to my work, I have the feeling of disguising myself. I admit it, it's hard, every morning I go to work, I have no choice, I have to work, otherwise I wouldn't do it, but I have a knot in my stomach. Of course I don't like it, when you go from a jilbab to a turban and a turtleneck, stuff that are way less large than a jilbab, I don't like it. (Camélia, a 30-year-old French-Algerian woman who wears either a jilbab or a turban)*
>
> *Indeed, I think I adapt myself a lot according to my interlocutors, according to moments, periods. I wouldn't go to my workplace in djellaba [2] with a black scarf around my head [...] I would rather wear a bandana instead. I would rather wear jeans. And I would rather go to the mosque with a skirt and a veil. You see? And so, it's because people who really know me know that I am a cha-meleon. (Fatou, 33-year-old French-Senegalese woman in hijab or turban)*

Such situations have led some veiled Muslim women not even to consider working in France, because they know in advance that they will suffer from major Islamo-phobic discrimination and significant injustice in finding a decent position that they deserve on the basis of their level of education and skills. The majority have even thought of leaving France to live in an Anglo-Saxon country (such as the United Kingdom or Canada), or a Middle Eastern country (such as Dubai) or even their country of origin (mainly in Africa), where they believe that it is easier to be accepted as a Muslim while working and partaking in any activities. This is the case for Kany and Myriam, who hope that the stigmatizing of veiled Muslim women in France stops and that they are offered the chance and the freedom to work like any other French citizen:

> *To be rejected at school, at work, I find it unfair. We, too, would like to have a professional activity, but it is getting more and more complicated [...] Once, I went to training centres to find a job and I didn't want to remove the veil. They told me that I had to take it off, or they wouldn't take me [...] So yes, I would like to leave France, I lived too many failures here. I realize that there is really a problem with this [finding a job]. Either in England or in my country of origin in Africa. (Kany, a 30-year-old French-Gambian woman in jilbab)*
>
> *Today, I would go anywhere. Honestly, I'm really looking for work everywhere and if I have an opportunity, yes, I'd go [...] at first... to Dubai. And secondly, I would love to find a job in England because they have a different vision. It's really not to attack France. I used to like France, but that's it, now, I want to leave. When I was looking for work (I have great professional experience, I was a manager), I said to myself: 'OK, I'm ready to be even an assistant'. Then, I asked myself: 'What has changed?'. Nothing, I just wear a headscarf. It's amazing, isn't it? I was even ready to be a maid, I would have done it just so as not to remove my headscarf. But then, I said: 'No'. (Myriam, a 31-year-old French-Algerian woman in hijab)*

The British interviewees do not share this desire to leave the country, and no one mentioned a desire to live in France. Ayesha (a 39-year-old British-Mauritian woman in *hijab*) does not even feel comfortable there on a simple tourist trip:

> *I have travelled a bit and I feel comfortable in Canada [...]. I've travelled to France many times and I would... I know you're French but I don't feel comfortable in France even before I started wearing the hijab. I think the racism is more obvious. (Ayesha)*

Veiled Muslim women express their concerns about Islamophobic discrimination differently, and it is interesting to explore the ways in which their clothing choices are used to mediate these concerns in everyday life. While many Parisian veiled women feel comfortable enough to wear the *abaya* or *jilbab* in their own neighbourhood, they prefer to wear a long skirt with a *hijab* in Central Paris (Najib & Hopkins, 2019). If these women vary their options, it is because they have noticed differences in their treatment by others and have found that they can manage to experience less Islamophobic hostility. Therefore, their Islamic clothing is worn strategically, and a notable dress code emerges clearly. This is indeed geographically located, especially in Paris where veiled bodies follow specific rules and behaviours.

For the British interviewees, the dress code is less problematic than in Paris; wearing a veil is certainly less of a problem than in France, where two laws prohibit the religious veil (the *niqab* in all public spaces, and the *hijab* in public schools and by representatives of public institutions). In contrast, UK laws protect religious diversity, and any form of religious discrimination is subject to strict judicial and legal sanctions. That said, several veiled British women explain that they also pay attention to the way that they dress in public areas and particularly on public transport. Their concerns are less about the *hijab* than the type of clothing they choose to wear. They tend to favour Western clothes and avoid traditional clothes, as explained below by Alia, Amelia and Nadia:

> I bought some [abayas] from Dubai and I was considering whether I should wear it or not, but I'm not sure, but they're so nice, so I feel like I should wear them. I do want to. It's just I always think in my head: 'Should I or should I not? What are people going to think of me?' I know I shouldn't think like that, but it's just I don't usually wear it, so it's a big thing for me. (Alia, a 26-year-old British-Pakistani woman who dons the hijab or the turban)
>
> If I'm travelling by Tube, never ever would I wear an abaya or a jilbab, it would be jeans or anything, just so that people don't feel... because sometimes your clothing really scares people [...] I'd always keep the hijab on but I would drop the abaya. (Amelia, a 34-year-old British-South Asian woman in hijab or jilbab)
>
> No. I would still wear my hijab. I might think twice, maybe, about wearing the jilbab in certain places. But in all honesty, my reasoning for that is more for security reasons. Because I'm thinking, if something happens and I have to run, I want to be dressed in a sort of way where it's practical and I can run quite easily. So maybe the jilbab, I'm slightly wary of where I would wear it, you know! But in terms of the hijab, no! (Nadia, a 35-year-old British-South Asian woman in hijab or jilbab)

Given the hypervisibility of the Islamic veil, veiled Muslim women have no choice but to develop strategies to make it less visible or more acceptable, depending on the context. They also experience phases in their life, for example being pushed to wear a simple turban when they are teenagers, then to wear a *hijab* as adults, then to take it off when they find a job, and so on. Thus, dressing and especially veiling are important embodied practices for Muslim women (Secor, 2002; Lewis, 2009; Dwyer, 2008; Gökariksel, 2012; Listerborn, 2015). The veil is not only a religious garment; it is also managed and controlled through space and time. Indeed, veiled Muslim women control in which spaces and places their religious identity can be seen, and also at what times and by whom.

Finally, all the forms of discrimination described in this sub-section (at work, in public spaces, on public transport, and so on) have a significant impact on the everyday behaviours of Muslim women, particularly on their fundamental freedom associated with choice of dress. The photos in Figure 4.1 show how a Muslim woman can dress differently. Muslim women and Muslims in general have various clothing styles that affect their relationships with others to a greater or lesser extent. This is why some Muslims also feel the need to prove that they are, above all, 'normal' people.

Towards more normalcy

In response to Islamophobia, both the Parisian and London respondents learned strategically to practise a type of self-regulation by performing normalcy. First and foremost, Muslim bodies (men and women) feel the need to put out a constant positive image of themselves. They want to show that they are 'nice people' – polite, sociable, funny and happy – to reassure people and ward off any Islamophobic situation. This is the case for Alya, a 21-year-old French student of Tunisian origin, who explains how her smile represents a suitable response to Islamophobic attitudes:

> *The fact that ones look at me in a bad way: 'Look at me badly, and I will smile at you'; it will never bother me [...] There are some people who will look at you in a weird way. I am really like that; my point is that we have to show them that we are not mean people. So I'm smiling at them [...] I smile, I smile, I smile. (Alya)*

This is also the case in the United Kingdom, where some veiled Muslim women want to show that they are not strange and dangerous people, especially on public transport. Here, many veiled Muslim women told me that they are emotionally affected by the fact that people choose not to sit next to them on the bus or Tube. This kind of comment was never made by the French participants. French people are not physically offended by veiled Muslim women to the point of wanting not to sit next to them. As explained in the previous chapter, French Islamophobia is more an institutional issue, while in the United Kingdom it is essentially an attitude problem that occurs in public spaces. Ayesha (a 39-year-old British-Mauritian woman in *hijab*) explains this painful feeling, and Laila (a British-Pakistani unveiled woman in her twenties) acknowledges the difference even though she does not even wear the *hijab* herself:

> *When I'm on the Tube and there're seats next to me, people don't come and sit next to me because I'm wearing the hijab, so that kind of makes me feel uncomfortable, because I'm just a human being just like them. (Ayesha)*
> *You get more looks on the Tube. Again, I know of friends who wear head-scarves and no one will sit next to them on the Tube. So, I guess, for them, probably their car is safer, but for me, I am pretty easy-going that way. (Laila)*

This attitude is more systematic towards women who wear the *niqab*. They are targeted more by such rejection. They have fully assimilated and normalized this attitude on public transport, forcing them to avoid it and even, for most of them, to stop using it.

To be exact, Muslim bodies do not want just to show that they are simply 'nice people'. They want to show that they are good integrated citizens. They want to show that they speak, look, eat and consume like any other French or British citizen. Whether French or British, men or women, the interviewees display certain behaviours that show them as well-integrated into society:

I would not wear that [the abaya] at work because I feel like I need to wear in a normalized way like our society is wearing, but with being modest at the same time so I could be like any other woman is. (Samia, a 31-year-old European resident in London with North African origin who wears the hijab)

I'm not taking out my Koran, nothing in Arabic. I'm really trying to get nothing out in Arabic on public transport. I try to take something that is not religious, because people are always curious, they always want to see what we read. You really have to take something neutral. At the end of the day, I read everything, so there is no problem. (Inès, a 20-year-old French-Tunisian woman in hijab)

At work... they're pretty shocked because I don't drink alcohol, they're pretty shocked. I've been working long enough and I've been through this for a long time. I understand that there are times when you have to omit the fact that you don't drink alcohol. But I did it spontaneously [...] it was an integration strategy. (Rachid, a French-Tunisian man in his forties who has a small beard)

Precisely to stand out, to speak real French, we don't speak in slang, we speak in French, we are in France, we speak in French no matter where you come from. Well, I don't want to talk about integration but you have to blend in with the masses, you see. (Habiba, a 29-year-old French-Moroccan woman in hijab)

[I sell] pies, cupcakes, very cute and very good things, and homemade with local products. I didn't want anyone to say that [my] pie shop is halal, you know! I really wanted everyone to be able to come. And why not, if they like the taste of halal, I can explain them more about Islam. (Habiba)

It's about changing the narrative, changing the perceptions. And the only way to do that is if we – I hate using this word – if we integrate, but integrate in the sense of you being part of discussions of how people view us [...] We need more people who are working in journalism, we need more people who work in policy and politics. (Laila, a British-Pakistani woman in her twenties who does not wear a headscarf)

As explained above, Samia seeks to integrate better by avoiding traditional clothes at work; Inès by avoiding reading Arabic books on public transport; Rachid by sometimes omitting that he does not drink alcohol; Habiba by eating and selling local products and by speaking proper French; and finally Laila by engaging in politics. They are not necessarily comfortable referring to the concept of 'integration', but this is what their behaviours reflect. That said, it is important to remember here that integration is a two-way process and to observe whether integration is really possible in the face of rejection, as Nida reminds us:

So, I don't know if people want to live with Muslims anymore, though. We don't mind living with others but it seems ironic that we get told that we don't integrate, but the irony of that is that we constantly get told, everywhere we go, 'You can't wear this. You can't look like this. You can't say this. You can't do this'. So, who's the one that's not integrating? (Nida, a British-Indian woman in her thirties who dons the hijab)

All told, Muslim bodies want to show that they are 'normal' people. They like acting in a specific way that is deemed normal and conforming. They believe that if they do not conform, they will be considered abnormal, unnatural, strange or, even worse, dangerous. Some interviewees have assimilated the idea that they have to prove themselves to people. They express it this way, using the term 'normal', as Wassim does:

> With regard to our behaviours whether at work, whether with our friends or people, we try to behave in a normal way, that's it! And I have the impression that it is more and more asked: a Muslim is asked to make efforts to be able to justify that he [or she] is normal. That is to say until proven otherwise, you are not normal. So, you have to show by your behaviours, by your attitudes that you are normal. That's what I noticed. Before I didn't have this view, I expressed myself normally. But now, I pay more attention to what I say. (Wassim, a French-Algerian man in his thirties who has a short beard)

They have taken up the idea that they must reassure people and show them that they are not too radical in their religious practice and that they are unrelated to any terrorist threat. Here, it is particularly the case for Muslim men, who are afraid of being listed in the so-called *Fiche S*[3] in France and the 'Prevent' programme[4] in the United Kingdom. This fear was much mentioned, and represents an important political concern for French and British Muslims. In this sense, Wassim adds:

> Now I've changed a lot. That is to say, there are things that I no longer say; there are websites that I will no longer visit when they are normal websites; for example, the word 'terrorist', I don't say it! [...] There are terms that I don't say because I always have the suspicion of being watched [...] Even in terms of places like parks, I go less and less because I still have the suspicion: 'What is he doing here? What is he preparing?'. (Wassim)

Even veiled Muslim women feel scared of being associated with terrorism. They can be perceived as supporters of Islamist terrorism because of their clothes or how they are portrayed in the media (Kwan, 2008; Runnymede Trust, 2017; Cohen & Tufail, 2017). More surprisingly, even non-veiled Muslim women can experience Islamophobic incidents that conflates their religion with terrorism. This is the case for Sirine and Laila, who have been perceived as terrorists:

> [At work, the day after the Paris attacks of 13 November 2015], my manager told me: 'Well listen, we can't keep you.' So I told her: 'Oh, why?' She did not answer [that day]. The next day, I told her: 'Can you tell me what is the reason why I cannot continue?' She looked at me and she said: 'It's because of terrorism.' Yeah, it is true that on the opposite sidewalk [to where I worked], there was the killing of the policeman. And then, I look at her and I told her: 'Why? Do you really think I'm a terrorist?' [...] And then I added: 'You know, the one who died on the opposite sidewalk [the policeman], his name is Ahmed, it is not François'. (Sirine, an Algerian woman in her fifties who resides in Paris)

> *As part of my job, I have to go to [the UK] Parliament, I work with poli-*
> *ticians. Every time I go through security, it is just very obvious that there is a*
> *lot of racial profiling [...] There was a group of us waiting to get in one of the*
> *entrances [...] We approached this police officer who was incredibly rude and*
> *insensitive, and starts yelling at me, and saying: 'You will cry, because one of*
> *you will say that you don't have any sharp objects in there [their bags], and*
> *then someone is going to blow themselves up.' At this point, he was looking at*
> *me, and he was just like: 'Some of you may not claim you are a terrorist, but*
> *there will be one of you that does.' And he was only looking at me. 'Terrorism is*
> *rife in this country, terrorism happens every day, there is an attack every day,*
> *you can't trust people, etc., etc.'. (Laila, a British-Pakistani woman in her*
> *twenties)*

For greater clarity, I asked Sirine and Laila what are the special markers that made possible this discrimination and association with terrorism. Sirine explained that her name shows that she is potentially a Muslim, and Laila indicates that the officer clearly 'identified' her as a potential terrorist by her skin colour. It is apparently not necessary to be a man or to wear a *hijab* to be associated with terrorism. There are many other markers such as name, skin colour, ethnicity... Tasnim (a British-Indian woman in her twenties wearing the *hijab* or *niqab*) stresses the importance for Muslims themselves to challenge all these negative stereotypes that overshadow the majority's ordinary and non-radical nature (Hopkins, 2004). In this sense, she teaches her children to be 'nice people' and to perform excellency because she does not want people to think that they are horrible:

> *We don't teach them [her children] to be horrible. We don't want them to grow*
> *up to be horrible people. We want them to be polite. We want them to mix in*
> *with everybody. We want them to be able to talk to people, go into the shop and*
> *say 'good morning' and 'good afternoon' and say 'hi' and 'thank you'. And have*
> *a quick conversation, 'I hope you're having a great day.' Being open to any sort*
> *of person in the community even if the person's walking by [...] You see an old*
> *lady, they [her children] can help her. An old man, they can help him. (Tasnim)*

To sum up, Muslim bodies have learned to live with Islamophobia and its normalization on a daily basis. Most of them even choose to ignore 'minor' acts, such as negative glances, bad comments, sighs and whispers (Najib & Hopkins, 2019). Everyday Islamophobia manifests as 'soft violence' in public space, and Muslims themselves trivialize certain actions. Some explain that they will not react each time someone looks at them negatively. Others explain that part of their religion, which is a highly resilient religion, teaches patience with people and to rely on God. But this type of 'passive' attitude does not mean that Muslim bodies who act as such cannot see and recognize oppression. Bell hooks (2000) and Philomena Essed (1991) explain that people who are genuinely oppressed know it, even if they do not react in definite terms, for example by engaging in activist associations or by formulating and filing complaints.

Several studies have shown that Muslim bodies are reluctant to report an Islamophobic incident (Carr, 2016; Paterson et al., 2019). This indicates that Muslims generally do not trust the police, believing that their experience will be downplayed. They also believe that the police are likely to act towards them unprofessionally and that their complaint will not be given serious considera- tion. They also think that society is generally blind to racism and Islamopho- bia, and that it is useless to put up a fight on a regular basis (Carr, 2016). Consequently, Islamophobic attitudes progressively become normal attitudes, within both society and Muslim populations. Here, there is a double process of normalization both to accept the threat and the deprivation of a fulfilling civic and social life. This soft violence certainly has a major long-term impact on Muslim bodies, potentially depriving them of full participation in public space (Noble & Poynting, 2008).

Finally, these strategies (invisibility and normalcy) that Parisians and Londoners create and adopt in public spaces can be very restrictive. For those who tire of pleasing everyone by hiding, conforming, smiling and ignoring, the only solution to Islamophobia is to become more vocal. Some Muslim bodies decide to become stronger and more assertive in order to defend their religious identity, which is constantly being attacked (Dwyer, 2008; Najib & Hopkins, 2019). Negative encounters laden with anti-Muslim feeling can, in turn, offer opportunities for change and a more positive attitude on the part of Muslim bodies. Some react by reaffirming their identity that is the most oppressed and by showing that 'Muslim is not dangerous' and in fact 'Muslim is beautiful', as other minorities have already achieved, such as the Black and the LGBT communities (Ramamurthy, 2013; Bloom & Martin, 2016; Herzog, 2011).

Ultimately, the behaviours and reactions of Muslim bodies to Islamophobia are fluid, leading them to react in public spaces in different ways with different people according to their mood, emotions, experiences, fears and concerns. However, knowing that their embodied religious identity can also be disrupted in more pri- vate and intimate spaces, without even sparing their own home, becomes increas- ingly disheartening.

Intimate Islamophobia in the interviewees' family space

After deploying strategies of invisibility or normalcy, Muslim bodies can also fall into a further strategy of intimacy by spending most of their time in the home space. Indeed, when it comes to Islamophobia and anti-Muslim hatred, a sig- nificant percentage of the 60 participants feel comfortable primarily at home (before any other space or place), in particular in Paris. Of the Parisian inter- viewees, 39% expressed this opinion, whether men or women, converts or not, against 18% in London where the interviewees are exclusively women wearing a veil (a *hijab, jilbab* or *niqab*) or not. Because of what they had experienced, they explained that they do not feel comfortable in public spaces and prefer to stay at home as often as possible for greater tranquillity, privacy and security of both body and mind. In Paris, one Tunisian woman (Emina) even explains that she feels at ease at home only once she locks the door of her apartment, while a French convert (Janna) clearly reports that she does not feel comfortable in her own home:

> *[I feel at home] in my apartment when I lock the door but otherwise nowhere. Being of Tunisian origin, I have no problem saying that [...] So for the moment, I feel nowhere at home, only when I lock [the door of] my apartment, unfortunately. (Emina, a Tunisian woman in her fifties who wears the hijab)*
>
> *Good question! Frankly I do not know. I don't know at all where I feel at home, because even in my own home I don't feel at home, with the new laws that have been passed... We had an unexpected visit from the police, not for us, for our neighbour who had done something stupid. They were looking for the neighbour all over the building, and they searched all the apartments, including mine. So we realize that even in our home, in fact, we are no longer at home. (Janna, a 31-year-old French convert who wears the hijab or jilbab)*

With the exception of these two women who scarcely feel at ease in their own apartments, the respondents feel most at home within the four walls of their house or apartment, surrounded by their relatives, as Ayesha sums up:

> *I feel at home when I'm at home at my house, and when I'm with my family, my nephew, my brother, my mum, sister-in-law. That's my home. (Ayesha, a 39-year-old British-Mauritian woman in hijab)*

These feelings and experiences are embedded in the domestic family space. Here, one can note not only how embodied experiences can be observed in this intimate space for some Muslims living in Paris and London but also how it represents a familiar and safe space. However, this domestic and family space can also represent a hostile environment where there is an everyday Islamophobia directly from family members.

Certain participants (a minority in the SAMA project) are affected by this more intimate-style Islamophobia. In this sense, Muslim bodies cannot find peace and serenity even at home, since their family represents for them a 'toxic entourage'. I met some interviewees in poor family circumstances who wanted to share their experiences. I did not expect them to tell me about incidents in their family, since none of the questions specifically targeted the family environment, yet they talked about it spontaneously. Throughout this qualitative research, I was able to learn more about intimate Islamophobia and its consequences for kinship relations.

First are converts (such as Emilie or Virginie) who have problems with their non-Muslim families due to their conversion to Islam. Several studies have shown that Muslim converts can be perceived as 'traitors' to their nation or race (OSF, 2011; Zempi & Chakraborti, 2014; Poole, 2002; Roy, 2013), and this feeling can be shared by their family regarding its religious heritage. Indeed, researchers have shown how difficult it is for converts to 'come out' as Muslims, especially with family members (Kose, 1996; Moosavi, 2011; Ramahi, 2020). Families who are against their conversion may indicate their disapproval through anti-Muslim sentiments and remarks. In addition, some Muslims (like Amine) have non-Muslim in-laws who believe what the media and politics broadcast about Islam and their followers. In both cases, the victims of intimate Islamophobia know that the anti-Muslim comments are primarily the result of ignorance and fear. The victims, for the most part, still wish to participate in family life, yet it remains a challenge since they choose to face this daily hostility in a non-confrontational way, trying to reassure the family:

> *So my family knows I'm wearing the veil but I take it off [when I visit them], and I feel like I have to do everything to show them that I haven't changed, even if it means multiplying my personality. But yeah, it's very complicated and it's a taboo [...] Because I realized that my conversion to Islam deeply hurt my family and that they live it very badly [...] But yes, yes, I know very well that I am a defect, that I have hurt my family enormously. My mother prays every day (since she is a Christian) for me to come back, to come back, yes, yes. (Emilie, a 29-year-old French convert in hijab)*
>
> *So when I started to wear the headscarf, I got a lot of grief, really a lot of it, first from my family. The first surprise was my family: 'What's going on with you, are you crazy or what? You are going to become a fundamentalist! What's wrong?' But, me, I haven't changed, it's just an accessory. So the first thing is with the family, it's extremely hard, they are the most difficult to convince that I haven't changed, it's always the same. I decided to wear the headscarf and I feel great with it, and it's really personal, it wasn't imposed. So the first difficulty is the family [...] that I have to continually reassure, reassure and reassure. (Virginie, a French convert in her thirties who dons the hijab)*
>
> *This is my point of view, I may be wrong, but we immediately see that everything that is negative revolves around Islam. If I can tell you for example a personal experience, it is that I have Catholic in-laws, and when I see their point of view, it is based on the media. They, without knowing Muslims directly or without knowing them much, they will judge by the media. So if the media say this is all about Islam, anything negative, insecurity and all that come from Islam, for example, they will agree with these judgements. (Amine, a 29-year-old French man of North African origin)*

Second are Muslims who started to practise their religion later in life or in a different way from their family. Their religious practices and especially the wearing of the veil have prompted Islamophobic attitudes within the family and sometimes even family breakdown. This is the case for Linda, who has a non-religious family that does not support her religious choice:

> *[My Muslim friends like me] encourage me to keep the headscarf, to stay who I am. But my family, no, because my family does not practise, and for them, the fact that I wear a headscarf also stigmatizes them. So that's also what causes the break. (Linda, a 31-year-old French-Algerian woman in hijab or turban)*

Likewise, Kany and Hawa, two French women of Sub-Saharan origin, see their families at odds with the headscarf and how they wear it:

> *It was mostly at the beginning when I started to put on the veil, it was badly taken in my family [...] There were even family friends who came to tell me: 'I don't want to see you with this anymore, take it off.' I did not listen to them. (Kany, a 30-year-old French-Gambian woman in jilbab)*
>
> *I was shocked and I didn't know what to do. I even asked those around me, my sister and my mother, if it was okay [that my boss told me to remove my veil], and what I should do. Honestly, I felt misunderstood because my mom [...] told me: 'If they ask you to take off the scarf, take it off'. My big sister told me: 'Me, I put on a bandana and I put on [hair] gel, you can do the same!' No! I was like: 'This – is – family?' We are supposed to help each other, we are supposed to have the same religion. (Hawa, a 27-year-old French-Malian woman in a hijab, and a turban at work)*

Indeed, in Sub-Saharan Africa, a veil is not necessarily the *hijab* which covers the hair as well as the neck, ears and chest; rather, it is the *boubou*,[5] which is more like a bandana or a simple turban, covering just the hair of many Black African women, including non-Muslims. Therefore, Black Muslim women who decide to wear the *hijab* in a more religious manner may suffer from Islamophobic comments directly from their families, wanting them to wear a veil according to their own cultural practice. In this sense, Cibo, a French Sub-Saharan woman, explains that in France, the veil worn by Black women seems to be more acceptable than that of other women. She comments:

> *I feel like when you are Black, it [the veil] is better perceived by people than when you're not Black. It's in our culture actually. Because we often have a turban or a boubou. So for them [French people in general], I feel like it's better, it's less shocking. When you see moms like that and everything, and they have their veil [their boubou], I feel like they [French people in general] are less shocked than when it's other people [non-Black wearers]. (Cibo, a woman in her thirties who wears the hijab)*

In France, Islamic clothes are worn by a broad and diverse constituency of Muslims, linked to their cultural background or the specific Islamic tradition they follow. Accordingly, Islamic clothes and particularly headscarves have become a powerful marker of social, cultural, political and spatial difference that may affect intimate space. Intimate Islamophobia focuses primarily on the headscarf in most of the examples studied here, and there are also some Muslim parents who do not want to see their daughters in *hijab* mainly for security reasons and concerns about their professional future, as Sakina explains:

> *Yeah, it was my mom actually who didn't want to at first. Overnight, I told her: 'Sorry, I leave you no longer the choice, I waited too long and for me, it is now or*

never, so tomorrow as soon as I go out, you will see me with the veil.' She was pissed off for two – three days and then, when she realized that we weren't going to stay mad all the time, she started to calm down and she told me that she was afraid. I was in high school at the time, so I was 15. For her, I think she was afraid of what could happen to me and she told me: 'Think about your studies, think about your studies', as if it was incompatible. I told her: 'Don't worry, it's not that tight, my brain still works'. (Sakina, a 21-year-old French-Algerian woman in hijab)

Finally, intimate Islamophobia undermines intimate relationships and has various motivations, all of which affect Muslim bodies' sense of self and belonging (Mansson McGinty, 2020). They become the object of anti-Muslim micro-aggressions (Islamophobic sentiments, comments, jokes, attitudes, etc.) within their own families, especially in France. In this study, intimate Islamophobia concerns exclusively French respondents certainly because, as explained previously, the wearing of the veil is less problematic in the United Kingdom than in France (due to the distinction in France between the secular public sphere and the private religious sphere) even among French Muslims. Some do not want to see the women in their family, in particular their daughters, signing their social and professional death warrant by wearing the *hijab*. Indeed, in France it is almost impossible for *hijabi* women to find a job (Rootham, 2014; Najib & Hopkins, 2019), and this reality obviously worries parents, pushing them to disapprove of wearing it. This is probably the reason why there are no examples of intimate Islamophobia among the British participants. It may also be because in London I did not meet any victims of Islamophobia who are converts or in an interfaith marriage. Only one woman told me that her boyfriend is of another religion (an Indian Sikh), but she did not report any particular issues with his family.

Conclusion

The findings discussed in this chapter on embodied and emotional Islamophobia show how this form of racism is alive and how it manifests itself in complex and multivariate ways. It triggers a whole range of behavioural tactics, ranging from self-invisibility and self-regulation of Muslim identities in public spaces to self-securitization in the enclosed space of the home. But in some cases, the domestic and family space – which is supposed to be a space where people feel at ease, especially for those who feel out of place in public (Noble, 2005) – also involves Islamophobia on a daily basis.

In our study, French participants seem more affected by intimate Islamophobia that undermines their kinship relations than the British, but also by strategies of invisibility to better hide the Islamic headscarf. In fact, veiled bodies are more affected by embodied and emotional Islamophobia, whether this occurs in the public or private space. Even though other bodies face an Islamophobic threat (Muslim men, unveiled Muslim women, etc.), veiled Muslim women, due to their great visibility and vulnerability, have to deal with several external factors caused by global negative representations from mainstream media and politics.

For example, the injunction to unveil in public spaces may push them into withdrawing from society and living a restricted life by becoming more involved in their family role (as wives and mothers). Indeed, these women represent the group most likely to be economically inactive and to elicit a significant level of scrutiny (Nagel, 2001; Ghani & Nagdee, 2019). They are either victims of discrimination in the labour market (Rootham, 2014) or channelled into traditional family roles of care and vocation (Bowlby & Lloyd-Evans, 2009; Farris, 2017). In the end, some of these women find no support for empowerment apart from themselves. Nevertheless, it is possible to find a few who have overcome all these obstacles to achieve an excellent level of education and an important professional position (Dale et al., 2002; Ahmad, 2001; Tyrer & Ahmad, 2006; Dwyer & Shah, 2009).

Importantly, Spatialized Islamophobia affects Muslim bodies and minds, while being connected to larger events of greater significance than mere personal practice or family quarrels. It reveals how the glocal approach of Islamophobia (ranging from worldwide Islamophobia to intimate Islamophobia) negatively impacts on Muslims' everyday lives, besides their aspirations and achievements. Muslim bodies negotiate their Muslimness at their own level according to different contexts, to the point where they are affected every day in terms of how they speak, work, dress, and so on.

By examining these various behaviours and practices, I was able to better understand through their bodies and minds how Muslims cope with anxiety, fear, risk and danger. Here, I have voluntarily provided many quotes from interviewees: who better to speak about Islamophobia than the victims themselves? They are the only ones entitled and empowered to expose their Islamophobic experiences and to explain the meaning of their behaviours. Thus, I would not have been able to write this fourth chapter if I had not met and talked with victims of Islamophobia both in France and the United Kingdom. Their individual and personal narratives have helped me not only to understand how they react in certain Islamophobic contexts but to recognize that Islamophobia is also present in more intimate and familiar spaces.

Accordingly, qualitative fieldwork was central to this study, and I am happy to have extended my skills beyond quantitative methods and to have learned new research techniques. Individual interviews and qualitative data allowed me to better examine and detail people's embodied and emotional experiences of Islamophobia, and thus to show that Spatialized Islamophobia is also embodied and emotional (Mansson McGinty, 2014, 2020; Listerborn, 2015; Najib & Hopkins, 2019; Fritzsche & Nelson, 2020). Indeed, it is important to apprehend the ways in which bodies interact with other bodies, emotions, spaces and events in their everyday lives, whether through interviews, observations or focus groups. Finally, after having analysed the various scales of Spatialized Islamophobia in Chapters 2, 3 and 4, as well as their interrelationships in Chapter 1, it is time in the next chapter to critically discuss and understand why the spatialized and multi-scalar nature of Islamophobia has been so little explored.

Notes

1 The *abaya* is a long dress worn by some Muslim women. It covers the entire body apart from the head, the hands and the feet. It can be more sophisticated than the *jilbab*, because it can be patterned, sparkly, etc. A headscarf is not included.
2 The *djellaba* is a long and wide dress, known as the traditional clothing of North Africa.
3 *Fiche S* means the S Card ('S' for State Security, in French *Sûreté de l'Etat*), an indicator used by law enforcement to flag and scrutinize an individual considered to be a serious threat to national security.
4 'Prevent' is a government counter-terrorism strategy aimed at preventing people from becoming terrorists or supporting terrorism. If a person is assessed as being a terrorism risk, they may be referred to the Home Office and assigned a mentor.
5 The *boubou* is the veil traditionally worn by Black African women. It is generally colourful and not necessarily religious. The term also refers to the long garment worn by both African men and women.

5 Conclusion: Towards a critical geography of Islamophobia

Introduction

This last chapter of the book goes beyond the objectives of the previous four which present the key points of the advanced concept of Spatialized Islamophobia. They detail the different geographies in which the various Islamophobias are embedded – in global geopolitics (Chapter 2), social geographies and ordinary urban encounters (Chapter 3) and emotional everyday geographies of bodies (Chapter 4) – but also their significant interrelationships (Chapter 1). Chapter 5 is the final stage of the analysis, which allows us to challenge the sense in which Spatialized Islamophobia is understood and studied. It does not seek to validate the spatialized and multi-scalar nature of Islamophobia, as in previous chapters, but to criticize why it has been so little explored in Islamophobia Studies and, especially, in Geography.

The discipline of Geography has been timid about studying Islamophobia, and geographers often refer to it as a form of systemic racism against Muslim populations without necessarily considering its spatialization. It is therefore more than important to refocus on 'space' and draw critical attention to the geographies of Islamophobia. It may seem, indeed, a little unusual to criticize a relatively new field of research in Geography, but my real criticism concerns this scarcity in itself: namely, why has it taken so long for the geographies of Islamophobia to be explored? Thus, this chapter questions the paucity of geographical approaches in two ways: 1) by showing the difficulty for many geographers in using a mixed-methods approach (essential to the study of Spatialized Islamophobia); and 2) by highlighting the whiteness of the discipline of Geography in the West and the role of racism within it.

The first critical point is to show that the combination of quantitative and qualitative approaches is not obvious in Geography (or even in other social sciences), and this may have hampered the study of Spatialized Islamophobia. Indeed, to better understand this phenomenon, which is based on a glocal process, it is essential to use both methods and to question why their combination is so rare in Geography. The way in which global Islamophobia is materialized and embedded in the conduct of everyday life shows how important it is to have an overview of this phenomenon. How can we grasp Spatialized Islamophobia

DOI: 10.4324/9781003019428-5

without being able to use multiple methods of analysis, ranging from the most quantitative to the most qualitative? In an attempt to integrate new methodological perspectives into the scholarship of geographies of Islamophobia, I allow myself to conduct here a first critical geography of Islamophobia by showing that most studies on Islamophobia and Muslim exclusion rely primarily on qualitative methodologies. It is necessary to criticize this lacuna of quantitative methods and call for a more mixed-methods approach (Hopkins, 2009; Najib & Teeple Hopkins, 2020) to the study of Spatialized Islamophobia. It is also necessary to better understand why so little quantitative geography is undertaken in such issues.

The second critical point concerns the dominant profile of geographers. This raises significant questions about the whiteness of the discipline and about its correspondingly greater difficulty in studying racial and religious oppressions from the point of view of the oppressed. Indeed, while focusing on how Islamophobia has been studied in Geography, I have realized that in fact this discipline has historically been (and still is) weak in discussing and exploring discrimination in general, and particularly discrimination in relation to 'religion' (Najib & Finlay, 2020) and 'race', compared to other social sciences such as Sociology. By spatializing topics such as Islamophobia, new perspectives and concerns regarding Islamophobia's sociological construction and manifestation could be garnered and then included in Geography. Geographers have the ability to do so because they have a close relationship with society and seek to challenge it for social change for the better.

In itself, this weakness, or I would even say blindness, in matters dealing with racial and religious discrimination and injustice constitutes a second critical geography of Islamophobia. This is explained in large part by the fact that Geography is a predominantly White discipline in Western universities, as noted by one of the reviewers and many other scholars (Kobayashi, 1994; Mahtani, 2014; Noxolo, 2017; Panelli, 2008; Mott & Cockayne, 2018). How can we demonstrate this phenomenon if few geographers benefit from 'strong objectivity' (Harding, 1998, 2004b) in this regard? We therefore need to hire more geographers from racial and religious minority groups to use this strong objectivity, not only to make a socio-spatial critique of the existing dominant relations of power and knowledge, but also to develop new programmes and areas of research in Geography to attain better knowledge.

From Critical Islamophobia Studies (CIS)...

Before presenting the critical geography of Islamophobia that I have identified, it is relevant to examine the various existing critiques of Islamophobia Studies. There is an emerging disciplinary field of Critical Islamophobia Studies (CIS) (Beshara, 2019; Allen, 2017), exemplified by many research works (notably the Islamophobia Research & Documentation Project (IRDP) at the University of California, Berkeley) in which academic-activists question the link between language and action; more precisely, between theory and praxis. These contributions are mainly from a sociological perspective, but there are many other ways to contribute to

CIS. Scholars from other social science disciplines (such as Politics, Law, Criminology, Economy, Psychology) offer articles on Islamophobia using critical thinking (Beydoun, 2016; Afshar et al., 2005; Garner & Selod, 2015; Ali, 2020; Mondon & Winter, 2017; Beshara, 2019; Lean & Esposito, 2012; Schwartz, 2010).

For example, the economic aspect of Islamophobia has been assessed and critically defined as a veritable industry, estimated at several hundred million dollars just in the US[1] (CAIR, 2015; Lean & Esposito, 2012; Schwartz, 2010; Bazian, 2015). The demonization of Muslims is a well-organized and well-funded business not only in the United States but also around the world. Similarly, the psychological perspective of Islamophobia has been explored (Inayat, 2007; Ad-Dab'bagh, 2017), particularly by Robert Beshara (2019) who offers a first critical approach of decolonial psychology to conceptualize everyday Islamophobia. He considers that the main objective of critical researchers working on CIS should be to respond intellectually to everyday Islamophobia through everyday resistance, using appropriate languages and actions. Finally, many theoretical resources are useful for CIS (such as theory and literature on decoloniality, postcolonialism, racism, feminism, liberalism, multiculturalism, universalism, and whiteness), and the essential reaction is radical engagement (essentially political) aiming to improve society.

Indeed, to be critical and radical means something. It means to grasp the root of the matter (Marx, 1844) and to engage actively in a revolutionary struggle that is rarely safe or pleasurable (hooks, 2000). It is essentially a question of intervening, theoretically and practically, in systems of domination and intersectional oppression (Soja, 2010). To be precise, if we are to eliminate Islamophobia (or any other form of oppression) effectively, we must develop a resistance movement that aims to have a radical transformative impact on society. This is all the more necessary as critiques and alternatives to anti-Muslim ideas are not facilitated. But we can no longer continue like this. We can no longer continue to see the progression of Islamophobia without doing anything. This progression is mainly due to a *laissez-faire* attitude and a comfortable silence that must end. Resistance engages us in revolutionary theory and praxis that radical and critical researchers place at the centre of the discussion. They seek to improve the world by denouncing and analysing its dysfunction and by envisioning and creating an adequate response to systems of domination.

Similarly, concerning the feminist struggle, Black feminist scholars have been able to identify their position of great marginality in order to better criticize the sexist and racist hegemony. Their contributions in the formation of feminist theory and praxis have been invaluable, enriching the discussion and enhancing the movement for authentic liberation and emancipation (Collins, 2002; hooks, 2000; Crenshaw, 1991; Carby, 1982; Spivak, 1988). Specifically, bell hooks (2000) explains that the experiences of people on the margins who suffer sexist and racist oppression can only be understood if they participate in the feminist and anti-racism movement 'as makers of theory and as leaders of action'. Thus, women have more to say to men about sexism, just as Blacks have more to say to Whites about racism, because men and Whites are usually not taught to recognize, respectively, their male and White privileges (McIntosh, 1988). In this context, the resistance to Islamophobia must definitely integrate the experiences of Muslims to enrich the critical contributions.

Simply on the basis of real anti-Muslim experiences, we can contradict one of the early critiques levelled at Islamophobia Studies: the one questioning the very existence of Islamophobia. This kind of narrative is often used by detractors who seek to delegitimize works on anti-Muslim hatred, whether these works are from academics, activists or policy-makers. But I do not work on something that does not exist or on a fiction. I have personally collected real statistical data from the Metropolitan Police and various associations. I have worked with actual associations fighting against Islamophobia. I have met real victims of Islamophobia in Paris and London, who have agreed to share with me their experiences of anti-Muslim discrimination.

In this sense, the following extract is from the evidence of a British participant, a woman of colour in *hijab*, whom I have renamed Khadija for anonymity. She agreed to talk to me about her particular experiences of anti-Muslim racism. Khadija is a remarkable woman as, in her profession, she plays a significant role in the fight against terrorist attacks on London soil. She also suffers from everyday Islamophobia due to her physical appearance:

> *In the last two years, [2] I've experienced so much racism and hatred towards me simply because I wear the hijab. I've been physically attacked on the Tube [...] On a weekly basis, I get called a terrorist all the time, whether it's on public transport or while I'm walking down the street [...] Since the 7/7 bombings, everything has been geared towards Muslims. It's all about Muslims, and we kind of get blamed for everything even though there are Muslims in our community, including myself who have worked with the law enforcement agencies to prevent Islamic – what they perceive as Islamic – terrorist attacks from occurring... I have worked at [a governmental institution]. I'm a counter-terrorism [officer] in my work, and I have assisted in several of the high-profile attacks that we've had and prevented three attacks that would have been worse than 9/11 in America. If I didn't care about my country, I wouldn't be doing that. (Khadija)*

It is striking how painful for Khadija is the contradiction between how she may be perceived and who she really is. She also feels that she will never be recognized for her true worth:

> *It feels like they [the government and the dominant population in general] want to ethnically cleanse Britain. They want it to be like predominantly White, and if we can't do certain things the way other nationalities in the community do, we're perceived as rebels because we don't want to learn the British way of life. But I'm British; I was born here, so that makes me 100% British in my eyes, so it's very difficult [...] I've experienced a lot of discrimination in the workplace as well [...] I haven't been given opportunities like my White counterparts to represent my department as much as others, even though I'm the most capable person to do so in my team. (Khadija)*

Some of her words refer largely to what Isakjee et al. (2020) term 'liberal violence', which seeks to elide both its violent nature and its racial underpinnings. We can note in these two excerpts that she refers to racism, White supremacy, postcolonialism and liberal contradictions. Indeed, if liberal societies see themselves as egalitarian, fair and peaceful, then Muslims should not suffer from increasingly normalized Islamophobic violence. Attempts by liberal societies to promote diversity were generally couched in a frame of assimilation that directly affects individuals. In this sense, Khadija's religious affiliation, clearly identified by her *hijab*, arouses Islamophobic attitudes, while she herself defends a set of important democratic values, such as diversity, freedom, feminism, and so on. Khadija robustly defends the British State against terrorism and thus participates in the protection of British citizens, yet she herself is not protected against violence and false representations related to her religion.

When I asked her what, in her opinion, were the special markers that made possible her experiences of Islamophobic discrimination, she almost instantly answered that it was her *hijab*, her skin colour and her gender. This calls into question the principles of liberal democracies, particularly through the lens of race and colonialism (also see Chapter 2). First, Khadija is a woman of colour from a foreign background and living in the West. Skin colour is fundamental in CIS, as Alexander shows when she says that it is impossible 'to separate Muslims from Black and Brown bodies' (2017: 15), or Moosavi (2015) or Selod (2015) when they explain that Islam is understood as a non-White religion. Within CIS, critical theories of race are needed to explain how a religious identity can elicit racial experiences (Meer & Modood, 2019; Sayyid & Vakil, 2010). The rejection of people seen as foreigners is a racial act and, in these postcolonial times, the legacy of colonialism continues to shape the relationships between different peoples (Moosavi, 2011). The West subordinated other societies by sending 'enlightened' White colonizers, who classified 'uncivilized Others' according to race and skin colour. For this reason, whiteness was associated with racial superiority and non-Whites with barbarism (Kumar, 2012). The White colonizers were able to dominate the colonized by making them believe that they were necessarily less civilized, less rational, less enlightened and even less human (Césaire, 2001; Fanon, 1967; Said, 1978). Put simply, the barbarian had to be civilized through violence in the name of Enlightenment, and this liberal contradiction continues to fuel contemporary racist thinking. Today, Western countries are still viewed as White nations by reinforcing the marginalization of postcolonial Others – including racialized 'Muslim Others'. This is what we call 'colonial amnesia' (Tyler, 2012) or 'racial amnesia' (Kapoor, 2013), whereby Western society forgets that historical subjects were rooted in racial processes, and therefore still maintains White supremacy, asserting that the privileges of the privileged are fair.

Second, Khadija is a religious Muslim person. In the predominantly secular context of our contemporary Western society, the mere fact of being a religious person can sometimes be deeply oppressive and challenging. In several European countries, the 20th century witnessed the establishment of secular states, leading to a significant decline in religious practices. Religion is taken into account less and less in the organization of European societies, and Islam is increasingly seen as a backward religion. As this view is reinforced by processes of racialization, it is also seen as foreign and dangerous (Said, 1978; Kumar, 2012; Scott, 2007). Therefore, the effects of these representations on anti-Muslim racism are unambiguous. These processes of racialization have been historically connected to European colonialism and question the very place of Muslims in European Western societies. Thus, the 'Muslim problem' has become a key issue in public debate in many Western democracies, as CIS analyses in depth.

Third, Khadija is a woman in *hijab*. In the West, veiled Muslim women are considered to be submissive and passive victims who must be rescued from their patriarchal religion and culture (Spivak, 1988; Mohanty, 2003; Fernando, 2009; Scott, 2007; Selod, 2015). They are also excluded from universal feminism by White feminism (Frankenberg, 1993), since they do not conform to the dominant norm, again in connection with colonial and imperial history (Macdonald, 2006; MacMaster, 2012). The defence of women's rights has often been used to stigmatize Islam and justify Islamophobic violence, mainly against women. In CIS, this type of mainstream Islamophobia against veiled Muslim women notably corresponds, according to Mondon & Winter (2017), to imperialist, paternalistic, racist and even sexist reassertions. Indeed, veiled Muslim women living in the West do not even have their freedom to dress as they wish respected or accepted.

Finally, these false representations that are used to justify Islamophobic violence through a liberal racist position (Mondon & Winter, 2017) hurt Khadija, as well as many other Muslims. Each of the Islamophobic tropes leads to the development of ideas that have direct consequences on actions and their legitimacy. Liberal Islamophobic violence is, in fact, more and more rampant, which ostensibly contradicts and betrays liberal values in terms of human rights and dignity (Isakjee et al., 2020). Thus, it is important for CIS to better understand how these dehumanizing representations of Muslims work and why they are so enduring. In other words, it is important for CIS to study the various ideological elements that make Islamophobia mainstream, normalized and acceptable.

CIS must also contribute to disturbing the current hegemony by building a counter-theory and praxis that are favourable to effective resistance. The counter-theory aims to show that these representations, for example those that hurt Khadija, are distorted representations of Islam and Muslims. Indeed, many studies have already shown that: 1) Muslims are, in fact, highly diverse individuals who are integrated into their societies (Halliday, 2003; Mills, 2009; Lewis, 2009); 2) religious Muslims are not associated with radicalization and extreme violence (Choudhury, 2007; Abbas & Siddique, 2012); and 3) Muslim women have their own agency (Bilge, 2010; Macdonald, 2006; Farris, 2017). Other counter-theories are possible for CIS, such as the rejection of the 'clash of civilizations', the recognition of good moderate Muslims, the development of inter-religious dialogue and the fight against White supremacy (Kumar, 2012; Ali, 2020).

As for counter-praxis, this is simply to do justice to all Western Muslims (not only them) because they, in particular, are still recovering from hundreds of years of imperialism, colonialism, postcolonialism, racism, discrimination and rejection

(Abbas, 2009). Islamophobic violence is a reflection of this, and still dictates the power relation between the powerful and the powerless. CIS analyses this relation and sheds light on broader debates on oppression and on the right to be different. The previous chapters have analysed this 'right to difference' (Esteves, 2019; Bowen, 2007) with a spatial lens, citing a 'right to appear' (Butler, 2011a), a 'right to the city' (Lefebvre, 1996), a 'right to centrality' (hooks, 1990, 2008; Young, 1990; Listerborn, 2015; Najib & Hopkins, 2019), a 'right to freedom' (Scott, 2007; Fernando, 2009), a 'right to control one's body' (hooks, 2000) and a 'right to dress' (Siraj, 2011; Dwyer, 2008). It is simply the right for Muslims to be themselves, without any injunction of total assimilation to the religious, cultural and racial norms of the dominant society. The study area of Spatialized Islamophobia may fuel scientific debate (and beyond) and its critical contribution, as detailed below, certainly counts among many others as an approach to CIS.

... to a critical geography of Islamophobia

As knowledge is under constant construction, CIS is able to welcome a new critical approach: one that is more geographical. The geographies of Islamophobia (Najib & Teeple Hopkins, 2020) seek to spatialize a thematic that has a more substantial tradition in social science disciplines other than Geography. Studies on Islamophobia and, more broadly, on race, religion and discrimination are extensive in Sociology and Political Sciences (McKee, 1993; Pager & Shepherd, 2008; Selod, 2015; Sayyid & Vakil, 2010; Alexander, 2017; Taras, 2013; Mondon & Winter, 2017; Ali, 2020; Kapoor, 2013; Moosavi, 2015), but the same cannot be said for Geography. Even though Islamophobia has always been spatialized and multi-scalar, it has not really been conceptualized as such, at least not in a concrete way. Negotiating and connecting the various spatial scales of social theorizing and political action is a rare and difficult task, even in Geography (Smith, 1992). Indeed, Geography fails to integrate the study of 'struggles' (whatever they are) at various scales and particularly to connect the scale of the body and the personal to the scale of the globe and the political-economic (Harvey, 2001).

Therefore, this book represents a fresh contribution to Geography, intrinsically positioning space and scale at the centre of the discussion. Many interests in Geography revolve around sociological phenomena, and by spatializing such phenomena, geographical expertise and new literature become available. By drawing attention to the role of space as a negotiating context (be it country, city, neighbourhood, home, mosque or workplace), the concept of Spatialized Islamophobia is better understood as both a powerful explanatory and exploratory research subject. Spatialized Islamophobia is explained as a process based on a glocal approach that explores the ways of connecting the microspace (of the body and 'mind') with the macrospace (of the globe). This geographical expertise and its criticism can certainly change the discussions that social geographers have conducted on race and religion for decades, since they have rarely been related to discrimination. To be precise, there are few geographical productions that show

how the construction of Muslims as 'threatening identities' relates to the defence of a White Europe (Hopkins, 2016) or to the ideas of liberal racism (Kong, 2009).

Most importantly, this chapter aims to better understand the various forces that influence the geographies of Islamophobia and to analyse them in a critical way. It is no longer a question of carefully clarifying the spatial meaning of Islamophobia, but rather of criticizing its weak implications and seeking to establish, for example, why there are so few spatial case studies or maps. This scarcity is inextricably social and cultural in nature (Harvey, 2001) and presupposes social and cultural ends (Pearson, 1957) revealing the lack of any means to achieve them. Consequently, it is necessary to examine how this critical geography of Islamophobia can be explained, especially in relation to the methodological traditions of Geography and geographers' predominant profile. Geography can be critical both methodologically and theoretically, as well as politically radical. It can challenge the systems of domination as well as the racist, patriarchal and capitalist structures of our contemporary society, and thus engage in critical thinking that is anti-racist, feminist and anti-capitalist. This critical thinking is in fact intersectional, and therefore also concerns Islamophobia.

Geography changes all the time and tackles new themes, because its very purposes have altered considerably over time. Initially, Geography was only about cartography (for the needs of the great explorers and navigators), then description (first seeking to make an inventory of the visible), then geopolitical (to serve governments, in particular to wage war), then theoretical and practical, whether quantitatively (with the help of statistical tools, mathematical models, etc.) or qualitatively (via interviews, focus groups, observation, etc.). Currently, the goal is no longer the inventory but the analysis of the relationship between people (and living things more generally) and their spatial environment, an analysis that differs from that of ethnologists and sociologists. The latter are only secondarily interested in the spatial realities in which people evolve. Indeed, those who work on Islamophobia have not (or only partially) addressed the question of its geography. Since geographers have opened up to Ethnography (for many reasons, such as the study of colonial geography), as well as to Sociology (especially under the influence of Marxism and Third Worldism) (Baud et al., 1997), geographers have all the tools to study topics related to domination and oppression.

Under these new influences, critical geography and radical geography emerged in the 1960s in the United States. These new geographies not only allow us to advance geographical thought and praxis inspired by great philosophers (Marx, De Certeau, Lefebvre, etc.), but also to question the political system of our liberal democracies as well as their capitalist economic model and their racist hegemony. But this new critical and radical current of thought represents limited areas of Geography, even today. Few geographers fully claim to be from critical geography and even fewer from radical geography, yet these currents should not be ignored in Geography because they continue to enrich and disrupt contemporary scientific debates, which are sometimes too conventional.

Critical and radical thought goes beyond mere explanation. It aims to explore the real causes that lie beyond appearances (Gintrac, 2012) and does not hesitate to take a stand against hegemonic theories. Critical geography and especially radical geography are inspired by various tendencies (anarchist, Marxist, feminist, anti-racist, anti-colonialist, etc.) going against more conventional Geography, which is often considered to be in the service of the dominant power and knowledge (Harvey, 1972, 1973; Bunge, 1971; Berg, 2009; Gintrac, 2012; Clerval, 2012). Many geographical works are positioned in this line, such as the influential works of David Harvey whose Marxist vision allowed him to associate Geography with capitalism, or the famous book by Yves Lacoste, *La géographie, ça sert, d'abord, à faire la guerre* (1976) (*Geography Is Primarily Used to Make War*), which denounces the use of Geography to secure more power. The creation of the geographical journal *Antipode* (in 1968) by activist-geographers (mainly from Clark University in Massachusetts, USA), frowned upon by traditional academia, also saw this current of critical and radical thought in Geography start to push at its edges (Peake & Sheppard, 2014; Paterson, 2014; Harvey, 2001). More recently, the journal *ACME* (*An International E-Journal for Critical Geographies*) has also been created to support more critical and radical perspectives; the activist and engaged dimension of this journal is clearly assumed, and it does not hesitate to publicly support militant movements like the Black Lives Matter movement for example.

All told, critical geography and especially radical geography seek not only to understand the socio-spatial problems of the everyday, but above all to find solutions for more social justice (Pickerill, 2016; Pande, 2020) while breaking free from the mainstream. Having said that, critical thinking and radical thinking should not be conflated: while radical thinking is systematically critical thinking, the inverse is not necessarily the case; critical thinking may not be radical (Gintrac, 2012). In this sense, Castree (2000) makes the distinction between critical geographers whom he considers to be better integrated into the academic world, and radical geographers whom he considers more marginalized. He denounces the institutionalization of critical geography and wonders whether it is possible in an institutional academic framework to denounce effectively the domination of capitalism, racism, sexism, Islamophobia, and so on. Even if the question is legitimate, this opposition between a more integrated institutionalized critical geography and a more marginal militant radical geography seems a little simplified and deserves further study.

Working on Islamophobia is already an important engagement. But this intellectual engagement is not enough to resolve this injustice without a collective engaged action. Critical and, especially, radical research is political (Pande, 2020), and research on Islamophobia naturally calls for a resistance that becomes praxis if the knowledge that it provides is to transform the oppressed consciousness (Freire, 1977). Thus, whether it is global Islamophobia or emotional Islamophobia, Spatialized Islamophobia in general needs spatialized resistance. This ubiquitous resistance (i.e. everywhere resistance) leads to the development of a critical rethinking of Geography. In this sense, Harvey (2001) explains that Geography hides devilish details that critical geography can combat to better counter dominant powers. According to him, the theoretical and practical ways of presenting geographical alternatives can directly influence emancipatory politics. He details:

Geography as we now know it was the bastard child of Enlightenment thought. It either remained hidden or, as with Kant, became the dark side of what the Enlightenment was supposed to be about. It is time to bring it actively into the light of day, legitimize it and recapture its emancipatory possibilities. That is, surely, the strongest of the 'strong ideas' that a critical geography can articulate at this difficult moment in our history.'

(Harvey, 2001: 233)

The critical geography of Islamophobia fits Harvey's explanation, because it, too, calls for political action, shows up the contradictions of liberal societies and questions the weakness of the geographical approach when studying Islamophobia in connection with its dominant traditions. Critical thinking is also necessarily more reflexive and raises the question of the positionality of researchers who, through their research, intend to participate in the fight against all forms of oppression. Certainly, those interested in my work on Islamophobia who have expressed the wish to contribute in various ways have overwhelmingly been geographers working on gender, sexuality, disability, race and racism. They have a more sympathetic understanding of issues relating to oppression, and correspondingly my work draws on feminist geography, postcolonial geography, and so on.

Finally, my positionality as a social and urban geographer, first quantitative and then qualitative, and as a female geographer from ethnic and religious minorities has helped me to build a critical geography of Islamophobia that focuses on the historical traditions of Geography, in particular on its usual methods and the power relations dominant in the discipline. First, the issue of the methods used in critical or radical geography has seldom been raised. These methods rely more on qualitative choices; which is also mainly the case for Islamophobia Studies in general and the geographies of Islamophobia in particular (Hopkins, 2009; Najib & Teeple Hopkins, 2020). As the various scales of Islamophobia have been little discussed, it is necessary to show the significance of apprehending Spatialized Islamophobia from a glocal approach, ranging from the largest scale to the finest. The various methods of analysis detailed in the previous chapters demonstrate that Spatialized Islamophobia has needed to be analysed both from a quantitative and qualitative perspective. But the difficulty of mixing these two types of methods in Geography is well known; there are various reasons (theoretical and political limitations, methodological divide, power influence, etc.), and it is certainly this weak communication between 'quantitative' and 'qualitative' geographers which can explain the belated demonstration of the spatialized process of Islamophobia.

Second, it is also essential to question the dominant power and knowledge of the academic context in which the geographies of Islamophobia will develop. Geography's lack of interest in Islamophobia and the importance of a strong objectivity in studying it can inspire critical observation of this discipline, finding it to be less inclusive than it should be, less egalitarian and less colourful. Indeed, the White dominance of this discipline has been noted in various topics (Kobayashi, 1994; Mahtani, 2014; Noxolo, 2017; Panelli, 2008; Mott & Cockayne, 2018) and may partly explain how Spatialized Islamophobia has been addressed in Geography.

If critical or radical geographies seek to transform the world for the better, then critical or radical geographers must first seek to change the world that they know best, namely the university and the knowledge it produces. No one should ever take for granted the truth of what they read; we should all seek to delve into and criticize theories, concepts and even contexts. We should not be afraid to criticize the context in which so many scholars have found their place. Myself, I am unafraid to expose the whiteness of the discipline of Geography and the role of racism within it because, while it may represent a risk to my future academic career, the most important thing for me is to promote my sense of justice in all kinds of knowledge, even those that make us less comfortable. In this sense, critical and especially radical geographies should not hesitate to address the complex challenge associated with how research engages the voices of marginalized groups (Pande, 2020). In the end, what is at stake in this book is a better understanding of Spatialized Islamophobia and a critical denunciation of its scarcity in Geography, and also – I would even say, above all – the 'establishment' of a more equitable world.

What has held back the study of Spatialized Islamophobia?

First critical point: The weakness of mixed-methods approaches in Geography

The study of Spatialized Islamophobia is a new field of study in critical or radical geography, which primarily seeks to place systems of domination at the centre of geographical thought. One of the geographers to have succeeded the most in dialectically connecting spaces with the social relations of domination is the British geographer David Harvey. He is the leading representative of English-speaking Marxist radical geography and one of the world's most cited thinkers in the social sciences (Clerval, 2011; Luxembourg, 2016). The philosophical and methodological evolution of this geographer interests me, particularly because it illustrates well my first critical argument on the geographies of Islamophobia which I develop here.

Harvey, like others, turned from quantitative geography to better apprehend qualitative realities. Early in his career in British universities, he naturally aligned his work with the dominant quantitative movement in Geography but quickly became aware of its theoretical and political limitations (Paterson, 2014). He left the United Kingdom in the 1970s for the United States, where he was struck by the pronounced economic exploitation of deprived neighbourhoods. He consequently felt the need to conduct more empirical and engaged work on the social and spatial relations of domination. Facing poverty and racism, as well as all the underlying problems, Harvey became increasingly aware of the inadequacies of the positivist geography of the 1960s and participated in the methodological debate of the time to support a more humanistic vision.

In some of Harvey's famous productions (1969, 1972, 1973), he advocates a major methodological change in quantitative geography to be more directly in contact with spaces and populations (Luxembourg, 2016). He developed a 'revolutionary theory' that, he says, only makes sense through 'revolutionary practice' (Paterson, 2014). Indeed, this revolutionary theory must above all have a materialist base and seek real 'humanizing social change' (Harvey, 1972: 122). Thereafter, Harvey's evolution continued towards more critical and radical geography, and the alternative that he proposes is drawn from Marxist thought since, for him, the goal of science is not only to understand the world as the positivist approach allows, but to change it notably via the Marxist approach. Starting from this new current of thought, Harvey shows the key place of space in the functioning of capitalism (Clerval, 2011), and he becomes the main leader of the 'breakthrough' from liberal to Marxist geography (Peet, 1977).

Many geographical reading grids are directly inspired by this Marxist vision of space, such as the centre-periphery urban model which emphasizes the dominant-dominated opposition (i.e. a centre concentrating various powers and dominating more or less marginalized peripheries (Burgess, 1925)), or Henri Lefebvre's (1996) right to the city that Harvey uses in his many books (2010, 2011, 2012). His works allow him to address broader concerns, such as hope, utopia, imperialism, the war on terror, and so on, as well as to raise the important issue of relevance in Geography. He concludes that it is impossible to develop relevant critical and, especially, radical thought from quantitative geography since it is too close to the State, which is often complicit in existing social problems and therefore in a process of manipulation and control to organize society's knowledge (Harvey, 1972). He questions the discipline of Geography as a whole and wants it to free itself from political, state and economic powers, and asks geographers (and researchers in general) to establish who they want to work for: politicians, directors of large companies or the public interest (Harvey, 2010). For critical and, above all, radical geographers like him, Geography must not only reappropriate its knowledge and tools for the purpose of social emancipation (Clerval, 2011), but also give voice to populations traditionally excluded from Western intellectual practices (Merrifield, 1995; Macdonald, 2006).

Harvey's trajectory and work illustrate well the difficult communication between 'quantitative' and 'qualitative' geographers, and allow us to question the philosophical and methodological framework capable of bringing together the dialectic of the global and the local. As I understand what Harvey and others have to say about quantitative geography, I think there is nothing fundamentally wrong with conducting quantitative research, although I have always felt the need to go further in my work and to 'zoom in' on more detailed phenomena that quantitative statistics cannot explain (Najib, 2013). Quantitative methods, assisted by modern computing tools, can be highly effective because they are capable of processing large volumes of data (such as census data). But even though I come from quantitative geography myself, I do not consider it to be a specific area of Geography but rather a set of techniques. For me, physical geography, social geography, urban geography, and so on, are areas of study, and we use quantitative and/or qualitative methods to answer specific questions relating to these areas.

Harvey considers that the quantitative movement can be interpreted partly 'as a rather shabby struggle for power and status within a disciplinary framework' (Harvey, 1972: 113) and shows a disturbing gap between increasingly sophisticated methods of analysis (e.g. factorial ecology, classification, statistical modelling, etc.) and the low relevance of the findings to tackling social problems robustly. While I partially understand his viewpoint, I do not agree with his view of an inherent opposition between critical thinking and quantitative analysis. On the contrary, I believe – like others (Barnes, 2009; Kwan, 2004; Openshaw, 1998; Sheppard, 2001; Poon, 2005) – that one can think critically and challenge the dominant powers by analysing quantitative data. I believe that we must simply consider the quantitative approach as a means of answering a given question and not as a science in itself, and we must especially distance ourselves from any institutional influence or political manipulation. Indeed, many researchers, whether from quantitative or even qualitative traditions, hired in the service of the government as specialists on a topic, have explained that they have seen their expertise elude them on the most sensitive and crucial points (Naudier & Simonet, 2011).

Most importantly, in my opinion, it is necessary to combine quantitative data with qualitative data in order to enrich the analysis and the findings. The previous chapters have shown the significance of quantitative statistics in the analysis of global, national and urban Islamophobias, just like the qualitative evidence to highlight infra-urban, embodied and emotional Islamophobias. There is therefore a critical need for mixed-methods approaches to the study of Spatialized Islamophobia. If I had not mixed the two, I would not have been able to develop this concept which explores both its macro and micro dimensions. Any social phenomenon visible on a large scale must be able to be compared to the behavioural realities observed on the ground, in actual situations. Thus, on the one hand I had to explore existing and accessible georeferenced quantitative databases and, on the other, to conduct individual interviews directly with victims of Islamophobia to better appreciate their internal point of view. Failure to explore both types of data in addition to the existing literature on mixed methods would not have helped me to conceptualize such a phenomenon, nor to observe, for example, whether or not the quantitative (i.e. recorded) and qualitative (i.e. perceived) geographies of Islamophobia correspond. Without such discussions, an important opportunity would have been lost to deploy the various existing methods to advance critical and radical geographical research and to effect changes to resolve religious discrimination.

We already know that in the social sciences, notably in Geography, mixing the two methods is not achieved so easily (Philip, 1998; Hodge, 1995; McKendrick, 1996; Brannen, 2017; Burgess, 1984), because researchers are generally trained in particular schools and departments, leading them to prefer one method over another (Brannen, 2017). The various existing methods are generally associated with specific epistemologies, and the discipline of Geography has evolved over a series of epistemological and methodological paradigms that have become norms and practices for many geographers (Goetz et al., 2009; Peet, 1998; Johnston, 2004). More accurately, it is common in Geography and in the scientific world in general to associate positivist traditions with quantitative methods, and more humanistic and postmodernist traditions with qualitative methods (Philip, 1998; Sheppard, 2001). In addition, each approach is associated with specific types of data collection and reasoning: the quantitative approach is generally seen as ungrounded and hypothetico-deductive, while the qualitative approach is instead seen as grounded and analytic-inductive (Brannen, 2017; Hammersley, 2017). These representations have created a divide between researchers (Hammersley, 2017). Nevertheless, some geographers (Sheppard, 2001; Goetz et al., 2009) have succeeded in somewhat reducing it, showing that the two logics go hand in hand and that their combination can bring complementary findings and develop new geographical concepts or even theories.

In the study of Islamophobia, the most-used methods remain qualitative whether in geographical studies or not. For example, many political scientists and sociologists have analysed Islamophobia in general terms derived from the discourses of global public debates, while most anthropologists, ethnologists and geographers have instead examined Islamophobia on a more individual level. The general tendency is therefore to adopt the qualitative approach. But when we do not consider the method as a science and when we do not feel dictated to by epistemology, this helps us to question the dominant methodological and epistemological framework. Indeed, developing and maintaining a stance independent of any scientific and academic influence may represent the starting point in the evolution of a study. The insufficiency of quantitative and mixed-methods approaches itself offers a critical examination of methodologies for researchers interested in investigating Islamophobia and Muslim exclusion. This paucity does not only concern the study of Islamophobia; it is also known in other fields and especially in gender and feminist studies (Peake, 2009). Qualitative methods are generally preferred by female scholars, as quantitative methods are mostly associated with a masculinist bias. Yet this is only a historically determined state of affairs that can be overcome (Brannen, 2017).

Consequently, it is essential to validate culturally the importance of mixing quantitative and qualitative methods in many social science disciplines and to build bridges between users of these two distinct approaches. The final objective is simply to promote the use of the two methods, or at least their exchange, in order to enrich each other and refine the results. The inability of a given researcher to use quantitative or qualitative methods should in no way be a matter of shame or contempt. Unfortunately, tensions are observed between 'quantitative' and 'qualitative' geographers, not only pushing some 'quantitative' geographers to believe that 'qualitative' geographers are conducting merely subjective research due to the lack of rigour in their methods, but also leading some 'qualitative' geographers to consider that 'quantitative' geographers cannot address socio-political issues (Philip, 1998). These simplistic representations are obviously false (Philip, 1998; Bennett, 1985; Gould, 1999; Lawson, 1995), and geographers should not be seen as either 'quantitative' or 'qualitative' geographers, but rather as geographers who use quantitative or qualitative methods.

Both methods have already proven their competence, and the combination has been found to strengthen geographical analysis as well as the influence of critical thought (Sheppard, 2001; Goetz et al., 2009). The most important thing is to satisfy the needs of a specific study rather than religiously following intellectual traditions. A taken-for-granted set of ideas is not fruitful in science: scientists must seek to explore each set of data and see how datasets complement or contradict one another. Combining quantitative and qualitative approaches is not necessarily efficient in all contexts, yet one usually enhances the other (Bryman, 1988; Brannan, 2017). This is why scientists should not be locked into these paradigms with their own terms, epistemologies and methodologies without going beyond the objectives that they had set in advance.

Finally, geographers can challenge themselves and put aside their differences to work together, make their own observations and analyse their own successes and failures, as each approach has its own strengths and weaknesses. For example, I can assert to other researchers that, as Islamophobia is still quantitatively understudied, it is very likely that the quantitative measures currently available cannot all capture the various aspects of the victims' everyday lives. Likewise, it is likely that certain qualitative realities can never be apprehended quantitatively. Sirin & Fine (2007) explain that quantitative and qualitative works should serve as a means of establishing a knowledge base on Islamophobia and, once this knowledge base grows, we can all gain new information and insights into this phenomenon.

The search for appropriate methods persists throughout this book and contributes to the development of a truly scientific science based on the desire to acquire more knowledge. Indeed, by analysing Islamophobia from the scale of the globe to that of the body (and mind), knowledge necessarily multiplies and provides findings that overlap in a relevant way. This 'real science' (Harding, 2004b: 137) is admittedly limited to processes controlled by methodological rules. But it must also be demystified because, in fact, this 'ideal science' takes place primarily in an environment that fosters the interests of a community of scientists largely made up of people who benefit – intentionally or not – from institutionalized racism (and sexism) and who, ultimately, must justify their findings to the institutions that recruit and pay them.

Second critical point: The whiteness of the discipline of Geography

As the university is a microcosm of contemporary society, this institution is also defined by racist, sexist and Islamophobic structures (Ahmed, 2012; Sian, 2017; Mirza, 2006; Patton, 2004; Fenton et al., 2000; Sue, 1993; Mir, 2014; Tyrer, 2003; Ghani & Nagdee, 2019; Mir & Sarroub, 2019). These various forms of oppression are intersectional, and the one that particularly interests me exposes the whiteness of the discipline of Geography (Bonnett, 1997; Kobayashi, 1994; Mahtani, 2014; Pulido, 2002; Panelli, 2008). This dominance is the result of wider systemic racist forces already in place that also affect other disciplines (Wong, 1994; Garner, 2007; Mills, 2004; Wakeling, 2007). To remedy this, Western universities regularly launch programmes promoting more diversity, yet the (neo)liberal tendency to increase the privileges of the privileged also involves academics. Geographers themselves, like the legal geographer David Delaney, have begun to show how Geography is a White discipline. He describes Geography 'as White an enterprise as Country and Western music, professional golf, or the Supreme Court of the United States' (Delaney, 2002: 11). This question of its whiteness – whether it is observed in the departments of universities, the knowledge they produce, the languages they use, their learning situations or their publishing practices – and the racism at its heart constitutes my second critical point on the geographies of Islamophobia.

First, Geography is a discipline that is unafraid to ask tough questions about itself (Sherman et al., 2005), and the question of its whiteness is one such question. This legitimate academic matter began to be studied in general terms in the early 1990s (Bonnett, 1997; Frankenberg, 1993; Kobayashi, 1994), and it was only later that geographers looked more specifically at the practices of discrimination and exclusion in academic spaces (Domosh, 2014; Faria & Mollett, 2016; Joshi et al., 2015; Mahtani, 2006; Abbott, 2006; McKittrick & Peake, 2005; Kobayashi, 2006). Bonnett (1997) explains that early studies on race and racism by historians of Anglo-American geography (such as Clarke et al., 1984; Jackson, 1987) did not even explore the issue of whiteness to the extent that White was not considered as a particular category of race. This practice reinforced racist patterns of thought, and this form of racism in the discipline began to be questioned. Indeed, racial minority groups are barely visible in many departments of Geography (Pulido, 2002)[3] and their absence reflects 'empty spaces' and 'silenced voices'. Some departments can even work primarily on racism and oppressed populations without ever including such people in their research teams. Yet, these groups deserve to occupy prime spaces and speak for themselves in Western universities (Kobayashi, 1994), because they are able to bring new knowledge to the field at odds with geographical traditions that are dominated by imperial and colonial ideologies and practices (Bonnett, 1997; Panelli, 2008). This White-dominated discipline disrupts the legitimacy of knowledge, and as with methods of analysis (discussed above), I am also critical of how Geography addresses particular topics.

Geographers must be able to criticize these ideologies by decolonizing geographical knowledge through a veritable decolonial critical approach (Panelli, 2008; Mahtani, 2006; Shaw et al., 2006). They should not underestimate the impact of these ideologies on their work and practices (Bonnett, 1997), but instead ask themselves why they are still active today (Noxolo, 2017; Noxolo et al., 2008). Similarly, fieldwork and conferences are not immune from criticism. They, too, are historically rooted in imperialist exploration and exclusionary practices that favour the maintenance of the whiteness of Geography (Abbot, 2006; Livingstone, 1993; Pulido, 2002; Oliver & Morris, 2020).

As a matter of fact, many aspects of Geography are influenced by its whiteness, and there are some invaluable contributions seeking to challenge them, such as the work of Black scholar Katherine McKittrick who brought new knowledge on 'Black geographies' into dialogue with traditional Geography. She filled the existing gaps by focusing on Black people's own concepts and methods (Black sense of place, plantation logics, etc.) (McKittrick, 2011). Other gaps exist, due to the low proportion of indigenous geographers and the rare recognition of their work (Noxolo, 2017), showing that the way in which geographers cite other geographers also depends on hegemonic structures that favour White Euro-American scholarship, especially the heteromasculine one (Mott & Cockayne, 2017;

Maddrell, 2015). This practice clearly reinforces the exclusion of non-White geographers from the scientific literature and contributes to silencing their voice further, instead of allowing them to exist in academia on their own terms. In exposing the negative impact of such practices on the lives of scholars of colour and on the scholarship on race and difference, Mahtani (2014) talks of 'toxic geographies' and denounces the norms and prejudices that underpin how geographical knowledge and institutions are reproduced (Panelli, 2008; Kobayashi & Peake, 2000). Even if this scholarship includes a decolonial and critical theorization, it still reproduces the same privileged spaces and practices and, therefore, contrasts with the actual marginalization of geographers of colour in Geography (Nayak, 2011).

This marginalization is not acceptable, and programmes calling for more diversity and inclusion will inevitably be less readily welcomed if their effect on the ground is very minimal. Like Mahtani (2014), I do not believe that the situation has improved, because there is no concrete action that has improved the situation of geographers of colour (especially in their recruitment). It is indeed increasingly tiring and demoralizing to read the statistics and testimonies revealing that they are the most marginalized, isolated and excluded from academic Geography (Desai, 2017; Tolia-Kelly, 2017; Mahtani, 2006; Panelli, 2008) in the absence of definite measures to change such injustice. Everyone deserves to be represented, heard, seen and valued. Representation is fundamental to what we expect from academia, and merely acknowledging the whiteness of Geography will not promote concrete change (Mahtani, 2014).

Each of us must criticize and reject this White privilege (even those who, directly or indirectly, benefit from it), because it normalizes an academic White supremacy. Failing to reject it amounts to accepting this 'systemic violence' (Young, 1990) which reveals a blatant gap between what the university says and what the university does. Above all, this struggle must be efficient, because otherwise it will be unable to escape degrading and even racist comments asserting that these racially minoritized researchers will only bring problems, be ungrateful, complain constantly, be too political and obtain their academic position only because of their skin colour (Ahmed, 2012).

The magnitude of the difficulties for these non-White researchers to access, for example, a permanent position or a professorship has no equivalent for their White counterparts (hooks, 2000; Hammou & Harchi, 2020). It is true that the job market is currently difficult, even for White academics, but they still benefit from pre-established tendencies that distinguish who must be saved from who must be left behind. These tendencies fragment the academic world, reproduce structural inequalities and perpetuate the processes of Otherness described throughout this book. Those who are perceived as universal, progressist and modern are those who need to be protected and helped to achieve a successful and secure career, while those who are perceived as marginal and who are used to being regarded as presenting multiple problems are those who constantly see their skills being questioned.

These ideologies must be combatted effectively, and one way is to change the recruiting system, which too often follows existing power forces and fosters affinities instead of skills (especially in France). Ultra-conformist academics follow dominant tendencies and can reproduce exclusionary phenomena without realizing it or seeing a problem. Many academics are more concerned with their own peace and career than justice, and may sometimes even join the ranks of the oppressors (Kobayashi, 1994). For all these reasons, it is more than necessary to review the members of scientific committees and, in particular, to include people who do not all look alike, who do not all think the same thing and who do not all behave in the same way. It is essential to put on each scientific committee at least one outsider or radical scholar capable of challenging the *Homo academicus* denounced by Bourdieu (1988) and dominant thought processes that may be racist, sexist, Islamophobic, and so on. This is how Sara Ahmed (2012) realized that the programmes calling for more diversity and equity in universities were, in fact, run by middle-aged White men who discussed what they used to eat for breakfast when they were together in Cambridge. Indeed, I do not believe that we can change the whiteness of an institution in this way. Bourdieu (1988) has shown that academia is not only a place of dialogue and debate, but also a sphere of power in which reputations and careers are defended or destroyed. Better monitoring of the practices of recruitment committees, as well as diversity and inclusion programmes, would be a good way to allow all academics, whoever they are, to work together without any domination whatsoever. Such engagement requires humility (Jaggar, 2004) to understand better the experiences of people of colour in academia, but also of women, queers, people with disabilities, Muslims, and so on.

Undeniably, academia is a sphere of power in which many forms of oppression are present, including Islamophobia (Mir, 2014; Tyrer, 2003; Ghani & Nagdee, 2019; Mir & Sarroub, 2019), which may be either latent (amateur and unconscious) or manifest (professional and conscious) (Bazian, 2014). There are also works that reveal the difficulties faced by Muslims (and mainly veiled women) in academia (Cole & Ahmadi, 2003; Speck, 1997; Mir, 2014; Barlow & Awan, 2016). Thus, it is important in this context to understand why these Muslim populations are marginalized in academia and what kind of knowledge this marginalization produces. With such marginalization and a predominantly White academic world, the study of Islamophobia encounters difficulty. The simple fact that Islamophobia has been a neglected theme in Geography is partly explained by the latter's continuing whiteness, which has impacted on our research and our way of doing Geography (Kobayashi, 1994; McKittrick & Peake, 2005). The geographical knowledge that it produces (sometimes even that which claims to be based on a decolonial approach) follows a form of White dominance that addresses topics primarily serving the interests of White academics and populations, and it is precisely this situated knowledge that demonstrates the difficulty of studying Spatialized Islamophobia.

Islamophobia is not one of the main concerns of mainstream geographers, and this is especially true since Muslim and non-White geographers are underrepresented in departments of Geography. How can they work on emerging concerns that directly affect them if their voices are constantly left out of academic conversations? These geographers benefit from strong objectivity in discussing Islamophobia, which refers to anti-Muslim racism and processes of racialization. Although it was not until the early 2000s that greater recognition of racialization took place in Geography (Kobayashi & Peake, 2000), Islamophobia includes the geographical scholarship of race and racism and may therefore relate to the reflexive approach of non-White geographers.

Undoubtedly, Geography needs the perspective of oppressed Others and their critical posture to bring fundamental knowledge to its creative development. In exploring Spatialized Islamophobia, the very absence of non-White or Muslim geographical thought necessitates a critical examination of the discipline of Geography and its ideologies, practices and knowledge. Mott & Cockayne (2018) go so far as to say that those who are 'othered' by disciplinary practices 'elevate a particular kind of knowing, of conveying knowledge, and of occupying intellectual space'. Indeed, as a Muslim geographer of colour, I am happy to participate in discussion based on my own interpretations and meanings of this phenomenon and to contribute to the literature on the geographies of Islamophobia. The spatialized and multi-scalar nature of Islamophobia is less explored than, for example, its processes of racialization in neighbouring disciplines such as Sociology, Anthropology and Religious Studies, and this relative dearth of geographical literature on Islamophobia does not help to enrich scientific debates. Geography is supposed to add value to what surrounds us, and if it shies away from topics other than those which are whiteness-focused, it may lose new perspectives of interdisciplinary discussions and research that would benefit of geographers and their students.

Accordingly, Geography as a discipline needs to be rethought so that it can include new concerns and interests related to oppressions generally studied first by other social science disciplines. The study of Spatialized Islamophobia has clearly been influenced by these other readings and may deviate from the pre-established norms of the discipline of Geography, but it was first guided by my identity and my desire to go beyond my comfort zone. Indeed, the teachings that I followed in quantitative geography in French academia do not really correspond to what I do now, however, researchers should be able to transcend their knowledge and not necessarily reproduce the knowledge of their 'masters', their supervisors and teachers. This is well explained by Harvey (2000: 254) when he argues that our ability to create new geographies is facilitated (or hindered) by: 1) where we can see geography from; 2) how far we can see geography; and 3) where we can learn geography from (as cited in McKittrick & Peake, 2005).

Ineluctably, more analyses, denunciations and testimonies are needed to contest the reproduction of this normative whiteness. Shifting its hegemonic practices (which are mainly unconscious) is not an easy task (Young, 1990) and requires a fundamental revolution within Geography. But such a challenge must be considered, because for too long it has 'traditionally included Western White men and excluded women, non-White communities, and non-Western geographical subjects; and the material and conceptual spatialization of difference' (McKittrick & Peake, 2005). This is what I have experienced personally throughout my studies and academic positions, and as I began the first pages of this book by using self-reflexivity, I wish to end it in the same manner by stating why, most of the time in academic Geography, I felt Othered.

Being a woman of colour in academia is not easy, and being a woman of colour in Geography is even worse. Being a quantitative geographer working on forms of oppression is not easy, and advocating the combination with more qualitative material is even harder. Being a Muslim academic working on Islamophobia is difficult, and even more so for those working in French universities where they are often treated with suspicion and seen as complicit in extremism. I have been able to work on Spatialized Islamophobia primarily because I moved from France to England, where the reflexive approach is more accepted and where studies on Islamophobia are more extensive. While I have been able to write this book, it has not been a straightforward journey, as explained at the start. Although I was socialized into White Geography and academic life, I have experienced much racist and Islamophobic discrimination and rejection, as follows.

First, as a young Maghrebi Muslim student from working-class and disadvantaged areas, I realized how the people in the various departments of Geography that I attended or visited in France (whether professors, students or administrative colleagues) did not look like me. Where were all the people like me? What were they doing? I quickly realized that it was not a normal situation to be so little represented, but as a student I could not denounce it, especially in a context where race and religion go unrecognized. Many French geographers found it suspicious that I was working on the issue of socio-spatial segregation, even though my work was based on existing literature. They wondered why I had chosen such a subject and questioned my objectives and purposes.

Second, I have seen White researchers working on forms of oppression help White students (mostly men) more effectively than students of colour. I have seen White men (sometimes in the highest positions in academia like professors, for example) using their power to harass and threaten women oppressively in precarious professional situations (temporary contracts, unemployment, etc.). I have seen, like hooks (2000), White feminist women fighting against male privilege in academia while discriminating against non-White women, sometimes in brutal ways. Like Ahmad (2003), I have seen White (and Black) feminists project preconceived attitudes that are as damaging as sexism and racism onto Muslim women. I have seen women of colour who adopt a Westernized feminist stance reaching a more secure academic position than those who have not rejected their cultural practice and still advocate for a more postcolonial version of feminism.

Third, I have seen White academics working on Islamophobia simply for fundraising opportunities, without having any real sympathy for Islam or any Muslim friends or acquaintances. I have seen Muslim and non-Muslim men working on Islamophobia who denigrate the work of Muslim women, sometimes using it to promote their own work on oppression and diversity. I have even seen White researchers working on Islamophobia who may themselves adopt Islamophobic attitudes. For example, some may show disapproval of those who do not drink alcoholic beverages at professional conferences or meetings. These minor incidents are in fact Islamophobic microaggressions, and they are deeply disturbing in researchers studying Islamophobia.

I have personally experienced all the above, like many other non-White geographers (Joshi et al., 2015) or more generally non-White scholars, who can be women or men, Muslim or not, veiled or not, from the West or not, and so on. Oppressed people are able to see and understand these aggressions (whether macro or micro) (hooks, 2000; Essed, 1991) even if they do not react immediately. In the end, the various examples described above show that oppressed academics can be oppressed in one place and privileged elsewhere. Therefore, we must all learn how to control our interests and opportunities (Harding, 2004a) and try not to reproduce the oppression that we have experienced in one specific area in another that we do not fully understand.

Clearly, my experience is not an isolated one, and it should not be treated as such. It is indeed a collective reality: racially minoritized academics (including Muslims) know, even in the 2010s to 2020s, what it is like to feel that there is no place for them in academia and that it is almost impossible for them to obtain a secure position. Some White academics even try to reassure non-White academics that the job market is tight for everyone, not just for racial minorities. But as they say this, they do not realize that non-White researchers may feel that before they can hope to find a job, White researchers must first secure theirs. White colleagues (obviously not exclusively) may subconsciously accept that it is almost normal to secure an academic position ahead of a non-White scholar, or to write a research project to support White students, or award them a (post)doctoral fellowship without going through formal and impartial recruitment, or find them a job before the end of their contract so that they do not face unemployment even for a month, or write a job offer that perfectly matches their profile. This (neo)liberal tendency to privilege the already privileged is easily observed in academia and seeks to make oppressed people 'feel that their situation is hopeless [and] that they can do nothing to break the pattern of domination' (hooks, 2000).

To be precise, this environment seeks to make oppressed people believe that they must accept this reality and refrain from applying for academic positions that are guaranteed to be given to people from dominant groups. Unfortunately, this strategy works. Even I, for example, have personally internalized the idea that I will never find a position in French academia because of who I am, what I study and how I study it. I also internalized the idea that I will not find a secure position in another country without going through an unfairly difficult journey to reach it. I know that some colleagues will roll their eyes to the heavens when reading this, but these statements should be taken seriously; otherwise it means that racially minoritized researchers cannot talk about racism within academia. The unequal treatment of racially minoritized researchers who work on research areas related to oppression (Spivak, 1988; Serrant-Green, 2002) represents a serious issue that must be addressed, since it can have important consequences on the functioning of academia (not only in France but also elsewhere). This important issue can sometimes lead racially minoritized researchers not to even apply, since they are certain that they have less chance of securing a position. This means that the academic system has already failed, as there is a problem in recruiting and attracting racialized minorities to specific positions (Carrington et al., 2000), just as with female researchers struggling to reach full professorship (Bower, 2012; Ginther & Kahn, 2009).

Acknowledging such dysfunction should not condemn us to victimization, or self-promotional complaint. Ahmed (2017) clearly explains that a complaint can be what we have to do because of how a space is occupied. If non-White academics were genuinely victimizing themselves, then they would not even have the necessary qualifications or professional experience to apply for academic positions. For most, the mere fact of having such a trajectory in itself represents a huge accomplishment. Those who victimize themselves do not work, complete a doctorate, publish articles, write books, and so on. Many academics of colour continue to work as if nothing was wrong. They are therefore anything but victims, and the injustices that they expose are anything but emotional complaints. On the contrary, these injustices are important facts and data to explore. Academics experiencing discrimination should not be afraid to speak out and stand up against the mainstream academic system.

As for me, I did not think that I would have achieved the things that I have achieved, but the more I was told I could not do something, the more I knew that I had to do it (for more representation, empowerment, hope, etc.). I readily admit that doing research and working in academia while criticizing it can seem a little contradictory and uncomfortable. But then, what should we do? Should we stop studying or applying? Should we accept that non-White and Muslim scholars only secure academic positions with a certain degree of luck? We have the right to ask ourselves these questions. We have the right to wonder if there are more non-Muslim (or ex-Muslim) academics working on Islamophobia and Islam in general in Western universities than Muslim researchers, as well as whether we prefer to hire non-religious Muslims to religious Muslims who would systematically have 'Islamist' ideas (Tyrer, 2003). Visible Muslims (such as veiled Muslim women, for example) undoubtedly have more difficulty in finding academic roles than non-visible Muslim researchers (and in the French case, it is even impossible). It is easier to create doubts about researchers in the category of 'Muslims', and even easier in the category of 'religious Muslims', because they are already under general surveillance and suspicion (Moors, 2019) and are usually stereotyped as representing radicalization. Therefore, their knowledge will be scrutinized in particular ways, and their non-inclusion, which may be quite obvious at certain moments of academic socialization, can have a direct impact on future recruitment.

In the end, it makes no sense to me to portray the phenomenon of Spatialized Islamophobia that occurs everywhere, without exposing it in academia. I promised myself, as someone who fights against all forms of injustice, that I will not leave the academic world without denouncing what I have experienced and witnessed with my own eyes. We must be able to say that the conformism of academia is distressing and unfair. If our testimonies can help to change the situation of future generations of researchers and especially geographers, then it is worthy. Our testimonies are not intended to discourage young students from racial and religious minorities from pursuing study or a career in academia. It aims to improve their situation or, at a minimum, simply to make them better understand the world in which they are to set foot, namely the same world (racist, Islamophobic, etc.) in which they operate on a daily basis. Even though I find it unfair to regard people who testify and expose their experience of discrimination as ungrateful and complaining, I could not silence my academic voice and I had to speak out against the injustice I was facing. I think it is essential to be able to criticize a situation, to alert, to denounce, to disturb and to value our personal academic experiences.

We need to be more ambitious about what our experiences can inspire in society. Criticizing something does not mean not liking it. I may have a serious problem with how some aspects of Western universities work, but I deeply love scientific research as well as challenging the world. Academia has to become better in the future, and Geography must become 'beautiful' and emancipate itself from everything that chains it to statism, capitalism, gender domination, racial oppression and imperialism (Springer, 2016). Towards that goal, I am unafraid to criticize academia while seeking to be integrated into it. Just because I do not look like most academics does not mean that I cannot be integrated into academia. I do not have to look like them, as bell hooks' mother reminded her. She told her that she could take useful knowledge from the mainstream group, but not participate in this knowledge to the point where it would lead to estrangement, alienation and, worse, assimilation (hooks, 2004). This is also what I believe: I will never look like them. I will always be this French-Moroccan Muslim woman of colour from working-class and disadvantaged areas.

Not looking like the mainstream is my strength, especially for academic purposes. It has allowed me to develop critical and radical thinking and thus go beyond what the mainstream system had planned for me. It does not matter if I am discriminated against because of my skin colour, my gender, my race, my religion, or if my name is scratched or, even, if someone tries to harass and intimidate me. I will always speak my truths, always be on the side of the oppressed and always 'think like a geographer', whether as an academic or not. Ultimately, discrimination is really not foreign to me, to the point where I am able to detect and distinguish at first glance between injustice and constructive criticism (as do other minorities (Essed, 1991; Moosavi, 2015)). Finally, my experience of academic discrimination has also strongly helped me, through this book, to contribute to Islamophobia Studies in general and to the geographies of Islamophobia in particular.

Conclusion

Faced with the Spatialized Islamophobia that is rampant all over the world, in political institutions, in the media, in the workplace, on the streets, in home neighbourhoods, in schools, and so on, I could not escape facing such injustice which also occurs in universities. Spatialized Islamophobia first targets postcolonial subjects by excluding them from any space, even spaces supposedly opposed to all forms of oppression. Many academics of colour, especially female geographers, have left academia (Monk & Hanson, 1982; Monk, 2006; Peake & Sheppard, 2014; Maddrell, 2012; Domosh, 2014) in the realization that it was much too difficult to make their point. This will certainly happen to me, too, but I am happy to have been able to develop this concept of Spatialized Islamophobia beforehand and thus show that Islamophobia is a spatialized process that occurs at various scales (Najib & Teeple Hopkins, 2020).

Specifically, this book provides a precise definition, and I wish to remind us of it, here in the conclusion: **Spatialized Islamophobia is a spatially pervasive form of anti-Muslim racism that occurs at various interrelated spatial scales (globe, nation, urban, neighbourhood and body (and mind)) and whose contours, effects, intensity and functioning vary accordingly.** Indeed, Spatialized Islamophobia is all of these: it is a kaleidoscope of the global, the transnational, the macro-regional, the national, the regional, the provincial, the metropolitan, the urban, the rural, the infra-urban, the local, the personal and the emotional, which are all in flux and interrelated. These spatial scales are indeed intimately interdependent and overlap in a very relevant way.

Spatialized Islamophobia undoubtedly contributes to debates in human geography at the intersection of social and cultural geography, urban geography, feminist geography, postcolonial geography, political geography, geographies of race and religion, and geopolitics. The geographies of Islamophobia highlight many important geographies, such as the geographies of Muslim identities and Islam; geographies of hope and contestation; geographies of residential segregation; geographies of deprivation and exclusion; geographies of home neighbourhood; geographies of perceived risk; geographies of fear, safety and belonging; geographies of embodiment, emotions and intimacy; and geographies of veiled Muslim women's bodies. They offer the opportunity to examine them critically (by questioning methods, theories, concepts, epistemologies, etc.). Other geographies could have been detailed here, such as rural geographies or youth geographies; for example, there is an interesting literature explaining how Muslims can feel 'out of place' in rural landscapes (Kinsman, 1995; Neal & Agyeman, 2006) and another covering the impacts of Islamophobia on spaces occupied by children and teenagers (Elkassem et al., 2018; Younus & Mian, 2018; Sirin & Fine, 2007). Future research can expand on the spatialized and multi-scalar nature of Islamophobia to explore these other spaces.

I hope, specifically, that this book will benefit all its readers and that they will rely on this concept of Spatialized Islamophobia as a theoretical anchor to explore several themes, whether for future geographical publications or not. Geographers could use this spatialized definition of Islamophobia, as could other social scientists who wish to give a spatial meaning to Islamophobia, considering the relationship between the globe and the body (and mind). Social scientists may eventually recognize and incorporate into their work on Islamophobia a geographical element by adding to their existing definitions that of Spatialized Islamophobia.

In conclusion, this book can give fresh impetus to scientific dialogue, because research never ends, and researchers continue to study various topics with their own eyes, as I have done here. Importantly, Spatialized Islamophobia should not be viewed as an issue that concerns only geographers or academics working on Islamophobia. It concerns everyone: academics, politicians, activists, policy-makers, Muslims and other minority groups, as well as the dominant part of society. Even if Spatialized Islamophobia objectively serves the interests of a specific ruling class, we are all concerned, and therefore we must all resist by building a liberation movement to end this type of oppression that history has already condemned.

Spatialized Islamophobia shows that Islamophobia is everywhere and, in this specific case, our goal as academics is no longer just to understand how the world works in space and time, but to improve it. This battle has chosen the Muslim populations who are in a position where they have no choice but to defend their religious identity, constantly under attack. They would certainly prefer that we talk less about their religion and more about the real added value that they bring, but oppression necessarily calls for resistance. Consequently, Spatialized Islamophobia (i.e. everywhere Islamophobia) needs spatialized resistance (i.e. everywhere resistance), just as everyday Islamophobia and racism need everyday resistance (Beshara, 2019; Dunn & Kamp, 2009). This resistance must not only be internationalized (from France to China, from the United States to Turkey, from Australia to Myanmar, etc.), but also localized (in mosques, schools, town halls, streets, etc.).

This world is such an uncertain place that it faces many threats (whether global or local), such as the terrorist and populist threat with the rise of extremism, the health threat with the spread of new viruses, the economic threat exposing new and rapid poverty, the climate and environmental threat degrading our planet, the legal threat reducing our basic rights, and so on. In the face of these threats, it is crucial to work together to defend our true democratic values of equality and justice for all, and not just for the privileged. Above all, we should not be afraid to stand up against those who have chosen to be on the side of inequality, injustice and oppression. We must especially continue to investigate the evolution of all types of hatred and to decide which strategy to prioritize. Should we focus more on prevention or awareness; should we raise awareness globally or more locally; should we work on policies or on political and media discourse; should we work on a better knowledge of our legal rights or more on ourselves and self-assertion?

In seeking to act on and modify the world, we discover ourselves and at the same time change our own nature (Marx, 1967: 173, as cited in Harvey, 2001). Most of the time, we are our own worst opponents in terms of making a difference and, as a result, we develop fears, barriers and divisions. But we must not become strangers to each other because we are not. We must accept and celebrate our differences rather than make them problematic, and thus enrich each other if we want to evolve in plural, free and respectful societies. At the end of the day, we only have one Earth, and we have no choice but to share it: so let's share it all together in real social justice... which means even against our own interests and privileges!

Notes

1 In the case of the US, between 2008 and 2013, over $200 million was spent to promote Islamophobic hatred, according to the Council on American-Islamic Relations (CAIR, 2015), although in 2016 Muslims represented only 1.1% of the total population (Lipka, 2017).
2 The interview took place on 7 February 2018.
3 Pulido (2002) shows that over 90% of Geography department members are White in the United States, and McKittrick and Peake (2005) explain that this comment could also apply to departments in Canada, New Zealand, Australia, Britain and Europe in general.

Appendix 1

References

Abbas, M. S. (2019). 'I grew a beard and my dad flipped out!' Co-option of British Muslim parents in countering 'extremism' within their families in Bradford and Leeds. *Journal of Ethnic and Migration Studies*, 45(9): 1458–1476.

Abbas, T. (2009). After 7/7: Challenging the dominant hegemony. In R. Phillips (ed.), *Muslim Spaces of Hope*, pp. 252–265. Zed Books.

Abbas, T., & Siddique, A. (2012). Perceptions of the processes of radicalisation and de-radicalisation among British-South Asian Muslims in a post-industrial city. *Social Identities*, 18(1): 119–134.

Abbott, D. (2006). Disrupting the 'whiteness' of fieldwork in geography. *Singapore Journal of Tropical Geography*, 27(3): 326–341.

Abdelkader, D. (2000). *Social Justice in Islam*. International Institute of Islamic Thought.

Abu-Lughod, L. (2002). Do Muslim women really need saving? Anthropological reflections on cultural relativism and its others. *American Anthropologist*, 104: 783–790.

Ad-Dab'bagh, Y. (2017). Islamophobia: Prejudice, the psychological skin of the self and large-group dynamics. *International Journal of Applied Psychoanalytic Studies*, 14(3): 173–182.

Afshar, H., Aitken, R., & Franks, M. (2005). Feminisms, Islamophobia and identities. *Political Studies*, 53(2): 262–283.

Ahmad, F. (2001). Modern tradition? British Muslim women and academic achievement, *Gender and Education*, 13(2): 137–152.

Ahmad, F. (2003). Still 'in progress?' – Methodological dilemmas, tensions and contradictions in theorizing South Asian Muslim women. In N. Puwar & P. Raghuram (eds), *South Asian Women in the Diaspora*, pp. 43–65. Berg.

Ahmed, L. (1992). *Women and Gender in Islam*. Yale University Press.

Ahmed, S. (2000). *Strange Encounters: Embodied others in post-coloniality*. Routledge.

Ahmed, S. (2004). *The Cultural Politics of Emotion*. Routledge.

Ahmed, S. (2012). *On Being Included: Racism and diversity in institutional life*. Duke University Press.

Ahmed, S. (2017). Complaint as diversity work. *Feminist Killjoys* blog. https://www.sarahahmed.com/complaint.

Aitchison, C., Hopkins, P., & Kwan, M-P. (2007). *Geographies of Muslim Identities*. Burlington.

Akhtar, S. (1989). *Be Careful with Muhammad! The Salman Rushdie affair*. Bellew.

Alexander C. (2017) Racing Islamophobia. In F. Elahi & O. Khan (eds), *Islamophobia: Still a challenge for us all*. Runnymede Trust, 13–16.

Ali, N. (2020). Seeing and unseeing Prevent's racialized borders. *Security Dialogue*, 51(6): 579–596.

Ali, Y. (2012). Shariah and Citizenship: How Islamophobia Is Creating a Second-Class Citizenry in America. *California Law Review*, 100: 27.

Allen, C. (2010). *Islamophobia*. Ashgate.

Allen, C. (2017). Ideological Islamophobia: Conception and function, 'normative truths' and 'new reality'. *Muslim Perspectives*, 2(2).

Allen, C. (2019). Governmental responses to Islamophobia in the UK; A two-decade retrospective. In I. Zempi & I. Awan (eds), *The Routledge International Handbook of Islamophobia*, pp. 397–407. Routledge.

Allen, C., & Nielsen, J. (2002). *Summary Report on Islamophobia in the EU15 after 11 September 2011*. Vienna: European Monitoring Centre for Racism and Xenophobia. http://fra.europa.eu/fraWebsite/attachements/Synthesis-report_en.pdf.

Allen, J., & Cochrane, A. (2014). The urban unbound: London's politics and the 2012 Olympic Games. *International Journal of Urban and Regional Research*, 38: 1609–1624.

Ameli, S. (2004). *The British Media and Muslim Representation: The ideology of demonisation* (vol. 6). Islamic Human Rights Commission.

Amin, A. (2002). Ethnicity and the multicultural city: Living with diversity. *Environment and Planning A: Economy and Space*, 34: 959–980.

Amin, A. (2012). *Land of Strangers*. Polity Press.

Amiraux V., & Simon, P. (2006). There are no minorities here: Cultures of scholarship and public debate on immigrants and integration in France. *International Journal of Comparative Sociology*, 47(3/4), 191–215.

Andersson, J., Vanderbeck, R. M., Valentine G., & Ward K. (2011). New York encounters: Religion, sexuality, and the city. *Environment and Planning A: Economy and Space*, 43: 618–633.

APPG. (2018). Islamophobia Defined: Report on the inquiry into a working definition of Islamophobia/anti-Muslim hatred. All-Party Parliamentary Group on British Muslims.

Archer, L. (2009). Race, face and masculinity: The identities and local geographies of Muslim boys. In P. Hopkins and R. Gale (eds), *Muslims in Britain*, pp. 74–91. Edinburgh University Press.

Ardener, S. (1981). *Women and Space: Ground rules and social maps*. Berg.

Avila, E. (2005). *Popular Culture in the Age of White Flight: Fear and fantasy in suburban Los Angeles*. University of California Press.

Bader, M. D., & Krysan, M. (2015). Community attraction and avoidance in what's race got to do with it? *Annals of the American Academy of Political and Social Science*, 660 (1): 261–281.

Balibar, E., & Wallerstein, I. (1991). *Race, Nation, Class: Ambiguous identities*. Verso.

Bangstad, S., & Bunzl, M. (2010). Anthropologists are talking about Islamophobia and Anti-Semitism in the New Europe. *Ethnos: Journal of Anthropology*, 75(2): 213–228.

Banton, M. (1955). *The Coloured Quarter*. Cape.

Barlow, C., & Awan, I. (2016). 'You need to be sorted out with a knife': The attempted online silencing of women and people of Muslim faith within academia. *Social Media + Society*, 2(4).

Barnes, T. J. (2009). 'Not only… but also': Quantitative and critical geography. *Professional Geographer*, 61(3): 292–300.

Bartolini, N., Chris, R., MacKian, S., & Pile, S. (2017). The place of the spirit: Modernity and the geographies of spirituality. *Progress in Human Geography*, 41(3): 338–354.

Baubérot, J. (2012). *La laïcité falsifiée*. La Découverte.

Baud, P., Bourgeat, S., & Bras, C. (1997). *Dictionnaire de Géographie*. Hatier.

Bayoumi, M. (2010). Being young, Muslim and American in Brooklyn. In L. Herrera, & A. Bayat (eds), *Being Young and Muslim: New cultural politics in the global South and North*, chap. 10. Oxford Press.

Bayrakli, E., & Hafez, F. (eds). (2019). *Islamophobia in Muslim Majority Societies*. Routledge.

Bazian, H. (2014). Latent and manifest Islamophobia: An inception of ideas. *Al Jazeera*, 15 April. https://www.aljazeera.com/opinions/2014/4/15/latent-and-manifest-islamophobia-an-inception-of-ideas/.

Bazian, H. (2015). The Islamophobia industry and the demonization of Palestine: Implications for American Studies. *American Quarterly*, 67(4): 1057–1066.

Bazian, H. (2019). 'Religion-building' and foreign policy. In E. Bayrakli, & F. Hafez. (eds), *Islamophobia in Muslim Majority Societies*, pp. 21–44. Routledge.

Bell, D. (2018). Europe's 'New Jews': France, Islamophobia, and Antisemitism in the era of mass migration. *Jewish History*, 32(1): 65–76.

Bennett, J. (2006). Flying while Muslim: Religious profiling? *Newsweek*, 21 November. http://www.newsweek.com/id/44711.

Bennett, R. J. (1985). Quantification and relevance. In R. Johnston, *The Future of Geography*, pp. 211–224. Methuen.

Berg, L.D. (2009). Critical human geography. In R. Kitchin, & N. Thrift (eds), *International Encyclopaedia of Human Geography*. Elsevier.

Berglund, J. (2009). *Teaching Islam: Islamic religious education at three Muslim schools in Sweden*. Doctoral dissertation, Uppsala University.

Bernardie-Tahir, N., & Schmoll, C. (2012). La voix des chercheur(-e)s et la parole du migrant. *Carnets de géographes*, 4.

Berry, B. (1964). Cities as systems within systems of cities. *Papers and Proceedings of the regional Science Association*, 13: 147–163.

Beshara, R. (2019). *Decolonial Psychoanalysis: Towards critical Islamophobia studies*. Routledge.

Beydoun, K. (2016). Donald Trump: The Islamophobia president. *Al Jazeera*, 9 November.https://www.aljazeera.com/indepth/opinion/2016/11/donald-trumpislamophobia-president-161109065355945.html.

Beydoun, K. (2018). *American Islamophobia: Understanding the roots and rise of fear*. University of California Press.

Bilge, S. (2010). Beyond subordination vs. resistance: An intersectional approach to the agency of veiled Muslim women. *Journal of Intercultural Studies*, 31(1): 9–28.

Blackwood, L. (2015). Policing airport spaces: The Muslim experience of scrutiny. *Policing: A Journal of Policy and Practice*, 9(3): 255–264.

Bleich, E. (2011). What is Islamophobia and how much is there? Theorizing and measuring an emerging comparative concept. *American Behavioral Scientist*, 55(12): 1581–1600.

Bloom, J., & Martin, W. Jr. (2016). *Black against Empire: The history and politics of the black panther party*. University of California Press.

Blunt, A. (2005). Cultural geography: Cultural geographies of home. *Progress in Human Geography*, 29(4): 505–515.

Boal, F. (1969). Territoriality on the Shankill-Falls Divide, Belfast. *Irish Geography*, 6: 130–150.

Bolognani, M. (2007). Community perceptions of moral education as a response to crime by young Pakistani Males in Bradford. *Journal of Moral Education*, 36(3): 357–369.

Bondi, L. (1998). Gender, class and urban space: public and private space in contemporary urban landscapes. *Urban Geography*, 19(3): 160–185.

Bondi, L., & Rose, D. (2003). Constructing gender, constructing the urban: a review of Anglo-American feminist urban geography. *Gender, Place & Culture*, 10(3): 229–245.

Bonn, S. A. (2012). The social construction of Iraqi folk devils: Post 9/11 framing by the G.W. Bush administration and US news media. In G. Morgan & S. Poynting (eds), *Global Islamophobia: Muslims and moral panic in the West*, pp. 83–99. Ashgate.

Bonnett, A. (1996). Anti-racism and the critique of 'White' identities. *New Community*, 22: 97–110.

Bonnett, A. (1997). Geography, 'race' and whiteness: Invisible traditions and current challenges. *Area*, 29(3): 193–199.

Bonnett, A. (2004). *The Idea of the West: Culture, Politics and History*. Palgrave Macmillan.

Bonnett, A., & Carrington, B. (2000). Fitting into categories or falling between them? Rethinking ethnic classification. *British Journal of Sociology of Education*, 21(4): 487–500.

Botterill, K., Hopkins, P., & Sanghera, G.S. (2019). Young people's everyday securities: Pre-emptive and pro-active strategies towards ontological security in Scotland. *Social & Cultural Geography*, 20(4): 465–484.

Bourdieu, P. (1988). *Homo academicus*. Stanford University Press.

Bourdieu, P. (2000). For a scholarship with commitment. *Profession*, 40–45.

Bourdieu, P. (2001). *Science de la science et réflexivité*. Raisons d'agir.

Bourdieu, P. (2004). *Esquisse pour une auto-analyse*. Raisons d'agir.

Bower, G. (2012). Gender and mentoring: A strategy for women to obtain full professorship. *Journal of Physical Education, Recreation & Dance*, 83(2): 6–12.

Bowen, J. R. (2007). *Why the French Don't Like Headscarves: Islam, the state, and public space*. Princeton University Press.

Bowlby, S., & Lloyd-Evans, S. (2009). You seem very westernised to me: Place, identity and Othering of Muslim workers in the UK Labour Market. In P. Hopkins & R. Gale (eds), *Muslims in Britain*, pp. 37–54. Edinburgh University Press.

Bowyer, B. (2009). The contextual determinants of whites' racial attitudes in England. *British Journal of Political Science*, 39: 559–586.

Brah, A., & Phoenix, A. (2004). Ain't I a woman?: Revisiting intersectionality. *Journal of International Women's Studies*, 5(3): 75–86.

Brannen, J. (2017) Combining qualitative and quantitative approaches: An overview. In *Mixing Methods: Qualitative and quantitative research*, pp. 3–37. Routledge.

Brennan, G. (2008). Homo economicus and homo politicus: An introduction. *Public Choice*, 137(3–4): 429–438.

Brenner, N., & Theodore, N. (2002). *Spaces of Neoliberalism: Urban restructuring in North America and Western Europe*. Blackwell.

Brice, K. M. A. (2009). Residential integration: Evidence from the UK census. In P. Hopkins & R. Gale (eds), *Muslims in Britain*, pp. 222–235. Edinburgh University Press.

Bryman, A. (1988). *Quantity and Quality in Social Research*. Unwin Hyman.

Bullock, K. (2005). *Muslim Women Activists in North America: Speaking for ourselves*. University of Texas Press.

Bunge, W. (1971). *Fitzgerald. Geography of a Revolution*. Schenkman.

Bunzl, M. (2007). *Anti-Semitism and Islamophobia: Hatreds old and new in Europe*. Prickly Paradigm Press.

Burgess, E. (1925). The growth of the city. In R. Park, E. Burgess, & R. McKenzie (eds), *The City*, pp. 37–44. University of Chicago Press.

Burgess, R. G. (1984). *In the Field: An introduction to field research.* George Allen & Unwin.

Burlet, L. (2017). L'Université Lyon 2 annule son colloque sur l'islamophobie. *Rue89Lyon,* 3 October.https://www.rue89lyon.fr/2017/10/03/luniversite-lyon-2-annule-colloque-lislamophobie/.

Butler, J. (2011a). Gender politics and the right to appear. Bodies in Alliance: The 2011 Mary Flexner lecture series. Bryn Mawr College.

Butler, J. (2011b). Bodies in alliance and the politics of the street. *European Institute for Progressive Cultural Policies,* 9: 1–29.

Cainkar, L. (2005). Space and place in the metropolis: Arabs and Muslims seeking safety. *City and Society,* 17(2): 181–209.

Cainkar, L. (2019). Islamophobia and the US ideological infrastructure of white supremacy. In I. Zempi & I. Awan (eds), *The Routledge International Handbook of Islamophobia,* pp. 239–251. Routledge.

CAIR. (2015). Unprecedented backlash against American Muslims after Paris attacks, Reports from the Council for American-Islamic Relations. https://www.cair.com/p resscenter/press-releases/13277-cair-reports-unprecedented-backlash-against-american-muslimsafter-paris-attacks.html.

Cantle Report. (2001). *Report of the Community Cohesion Review Team, Institute of Community Cohesion.* Report led by Ted Cantle, Home Office.

Carby, H. (1982). White women Listen! Black feminism and the boundaries of sisterhood. In R. Tavernier (ed.), *Centre for Contemporary Cultural Studies, The Empire Strikes Back: Race and racism in 70s Britain.* Hutchinson.

Carr, J. (2016). *Experiences of Islamophobia: Living with racism in the neoliberal era.* Routledge.

Carrington, B., Bonnett, A., Nayak, A., Skelton, C., Smith, F., Tomlin, R., Short, G., & Demaine, J. (2000). The recruitment of new teachers from minority ethnic groups. *International Studies in Sociology of Education,* 10(1): 3–22.

Cassiers, T., & Kesteloot, C. (2012). Socio-spatial inequalities and social cohesion in European cities. *Urban Studies,* 49(9): 1909–1924.

Castree, N. (2000). Professionalisation, activism and the university: Whither 'critical geography'? *Environment and Planning A: Economy and Space,* 3: 955–970.

Castree, N., Kitchin R., & Rogers A. (2013). *A Dictionary of Human Geography.* Oxford University Press.

CCIF. (2011). *Rapport annuel sur l'islamophobie en France en 2010.* CCIF, Collectif Contre l'Islamophobie en France (Collective against Islamophobia in France).

CCIF. (2016). *Rapport annuel sur l'islamophobie en France en 2015.* CCIF, Collectif Contre l'Islamophobie en France (Collective against Islamophobia in France).

Césaire, A. (2001 [1955]). *Discourse on Colonialism.* New York University Press.

Cesari, J. (2005). Ethnicity, Islam, and les banlieues: Confusing the issues. *Social Science Research Council,* 30.

Chao, E. (2015). The-Truth-About-Islam.Com: Ordinary theories of racism and cyber Islamophobia. *Critical Sociology,* 41(1): 57–75.

Charles, C. Z. (2003). The dynamics of racial residential segregation. *Annual Review of Sociology,* 29(1), 167–207.

Charmes, E. (2004). Les Gated Communities: Des ghettos de riches? *L'Espace géographique,* 2: 97–113.

Charmes, E. (2009). Pour une approche critique de la mixité sociale. Redistribuer les populations ou les ressources? *La Vie des idées.* http://www.laviedesidees.fr/Pour-une-approche-critiquede-la-mixite-sociale.html.

Cheruvallil-Contractor, S. (2012). *Muslim Women in Britain: De-mystifying the Muslimah.* Routledge.

Chignier-Riboulon, F. (2010). Les quartiers en difficultés, une question d'intégration. In G. Wackermann (ed.), *La France en villes*, pp. 185–191. Ellipses.

Chiseri-Strater, E. (1996). Turning in upon ourselves: Positionality, subjectivity, and reflexivity in case study and ethnographic research. In P. Mortensen & G. Kirsk (eds), *Ethics and Representation in Qualitative Studies of Literacy*, pp. 115–133. National Council of Teachers of English.

Choudhury, T. (2007). *The Role of Muslim Identity Politics in Radicalisation.* Department for Communities and Local Government.

Clarke, C., Ley, D., & Peach, C. (1984). *Geography and Ethnic Pluralism.* Allen and Unwin.

Clayton J. (2009). Thinking spatially: Towards an everyday understanding of inter-ethnic. *Social & Cultural Geography*, 10(4): 481–498.

Clerval, A. (2011). David Harvey et le matérialisme historico-géographique. *Espaces et sociétés*, 4: 173–185.

Clerval, A. (2012). Gentrification et droit à la ville. *La Revue des livres*, 5: 28–39.

Clerval, A. (2016). *Paris sans le peuple: la gentrification de la capitale.* La Découverte.

Cohen, B., & Tufail, W. (2017). Prevent and the normalization of Islamophobia. In S. Cohen (1972). *Folk Devils and Moral Panics: The creation of the mods and rockers.* MacGibbon and Key.

Cohen, S. (1972). *Folk Devils and Moral Panics: The creation of the mods and rockers.* MacGibbon and Key.

Cole, D., & Ahmadi, S. (2003). Perspectives and experiences of Muslim women who veil on college campuses. *Journal of College Student Development*, 44(1): 47–66.

Collignon, B. (2010). L'éthique et le terrain. *L'information géographique*, 74(1): 63–83.

Collins, P. H. (2002). *Black Feminist Thought: Knowledge, consciousness, and the politics of empowerment.* Routledge.

Copsey, N., Dack, J., Littler, M., & Feldman, M. (2013). *Anti-Muslim Hate Crime and the Far Right.* Centre for Fascist, Anti-Fascist and Post-Fascist Studies, Teesside University.

Crenshaw, K. (1991). Mapping the margins: Intersectionality, identity politics, and violence against women of color. *Stanford Law Review*, 43.

Dabashi, H. (2012). *Corpus anarchicum; Political protest, suicidal violence, and the making of the posthuman body.* Palgrave Macmillan.

Dale, A., Shaheen, N., Fieldhouse, E., & Kalra, V. (2002). Labour market prospects for Pakistani and Bangladeshi women. Work, *Employment & Society*, 16(1): 5–26.

Damasio, A. R., Grabowski, T. J., Bechara, A., Damasio, H., Ponto, L. L., Parvizi, J., & Hichwa, R. D. (2000). Subcortical and cortical brain activity during the feeling of self-generated emotions. *Nature Neuroscience*, 3(10): 1049–1056.

Danis, S. (2016). *La fantasmatique du grand remplacement dans le roman français contemporain (Renaud Camus, Eric Zemmour, Michel Houellebecq).* MA thesis, University of Montréal.

Datta, A. (2009). Making space for Muslims: Housing Bangladeshi families in East London. In R. Phillips (ed.), *Muslim Spaces of Hope*, pp. 120–136. Zed Books.

Davis, A. Y. (2016). *Freedom Is a Constant Struggle: Ferguson, Palestine, and the foundations of a movement.* Haymarket Books.

Day, K. (1999). Embassies and sanctuaries: Women's experiences of race and fear in public space. *Environment and Planning D: Society and Space*, 17: 307–328.

De Certeau, M. (1984). *The Practice of Everyday Life*. University of California Press.

De Galembert, C., & Belbah, M. (2004). Vertus heuristiques d'une recherche en tandem. La gestion publique de l'islam en France (enquête). *Terrain & Travaux*, 7(2): 127–145.

Delaney D. (2002). The space that race makes. *Professional Geographer*, 54: 6–14.

Delphy, C. (2006). Antisexisme ou antiracisme? Un faux dilemme. *Nouvelles Questions Féministes*, 26(1): 59–83.

Desai, V. (2017). Black and Minority Ethnic (BME) student and staff in contemporary British Geography. *Area*, 49(3): 320–323.

Desponds, D., & Bergel, P. (2017). Identifier les 'quartiers sensibles' dans les villes françaises: une quête sans cesse recommencée. De la directive HVS de 1977 à la nouvelle géographie prioritaire de 2014. *Itinéraires. Littérature, textes, cultures*, 2016-2013/2017.

Dikeç M. (2006). Badlands of the Republic? Revolts, the French state, and the question of banlieues, *Environment and Planning D: Society and Space*, 24: 159–163.

Dixon, D., & Marston, S. (2011). Introduction: Feminist engagements with geopolitics. *Gender, Place & Culture*, 18(4): 445–453.

Domosh, M. (2014, August 1). The More-than-Conference Conference, AAG President's Column. http://news.aag.org/2014/08/the-more-than-conference-conference/.

Dorlin, E. (2008). *Sexe, genre et sexualités. Introduction à la théorie féministe*. PUF.

Duncan, P. (2016). Europeans greatly overestimate Muslim population. *The Guardian*, 13 December.https://www.theguardian.com/society/datablog/2016/dec/13/europeans-massively-overestimate-muslim-population-poll-shows

Dunn, K., Atie, R., Mapedzahama, V., Ozalp, M., & Aydogan, A.F. (2015). *The Resilience and Ordinariness of Australian Muslims: Attitudes and experiences of Muslims report*. Western Sydney University and the Islamic Sciences and Research Academy.

Dunn, K., & Hopkins, P. (2016). The geographies of everyday Muslim life in the West. *Australian Geographer*, 47(3): 255–260.

Dunn, K. M. (2005). Repetitive and troubling discourses of nationalism in the local politics of mosque development in Sydney, Australia. *Environment and Planning D: Society and Space*, 23: 29–50.

Dunn, K. M., & Kamp A. (2009). The hopeful and exclusionary politics of Islam in Australia: Looking for alternative geographies of 'Western Islam'. In R. Phillips (ed.), *Muslim Spaces of Hope*, pp. 41–66. Zed Books.

Dunn, K. M., Klocker, N., & Salabay, T. (2007). Contemporary racism and Islamophobia in Australia. *Ethnicities*, 7: 564–589.

Dwyer, C. (1999). Veiled meanings: Young British Muslim women and the negotiation of differences. *Gender, Place & Culture*, 6(1): 5–26.

Dwyer, C. (2000). Negotiating diasporic identities: Young British South Asian Muslim women, *Women's International Studies Forum*, 23(4): 475–486.

Dwyer, C. (2008). The geographies of veiling: Muslim women in Britain. *Geography*, 93 (3): 140–147.

Dwyer, C., & Meyer, A. (1995). The institutionalisation of Islam in the Netherlands and in the UK: The case of Islamic schools. *Journal of Ethnic and Migration Studies*, 21(1): 37–54.

Dwyer, C., & Shah, B. (2009). Rethinking the identities of young British Pakistani Muslim women: Educational experiences and aspirations. In P. Hopkins and R. Gale (eds), *Muslims in Britain*, pp. 55–73. Edinburgh University Press.

Dwyer, C., Shah, B., & Sanghera, G. (2008). From cricket lover to terror suspect' – Challenging representations of young British Muslim men. *Gender, Place & Culture*, 15 (2): 117–136.

Dwyer, C., & Uberoi, V. (2009). British Muslims and 'community cohesion' debates. In R. Phillips (ed.), *Muslim Spaces of Hope*, pp. 201–221. Zed Books.

Ehrkamp, P. (2007). Beyond the mosque: Turkish immigrants and the practice and politics of Islam in Duisburg-Marxloh, Germany. In C. Aitchison, P. Hopkins, & M-P. Kwan (eds), *Geographies of Muslim Identities*, pp. 11–28. Ashgate.

Elkassem, S., Csiernik, R., Mantulak, A., Kayssi, G., Hussain, Y., Lambert, K., ... Choudhary A. (2018). Growing up Muslim: The impact of Islamophobia on children in a Canadian community. *Journal of Muslim Mental Health*, 12(1): 3–18.

Elshayyal, K. (2018). *Muslim Identity Politics: Islam, activism and equality in Britain*. I. B. Tauris.

El Zahed, S. (2019). Internalized Islamophobia: The making of Islam in the Egyptian media. In E. Bayrakli & F. Hafez (eds), *Islamophobia in Muslim Majority Societies*, pp. 137–160. Routledge.

Essed, P. (1991). *Understanding Everyday Racism: An interdisciplinary theory*. Sage.

Esteves, O. (2019). A historical perspective: secularism, 'white backlash' and Islamophobia in France. In I. Zempi & I. Awan (eds), *The Routledge International Handbook of Islamophobia*, pp. 99–109. Routledge.

Falah, G.W., & Nagel, C. (2005). *Geographies of Muslim Women: Gender, religion and space*. Guilford.

Fanon, F. (1967). *Black Skin White Masks*. Grove.

Faria, C., & Mollett, S. (2016). Critical feminist reflexivity and the politics of whiteness in the 'field'. *Gender, Place & Culture*, 23(1): 79–93.

Farris, S. (2017). *The Name of Women's Rights: The rise of femonationalism*. Duke University Press.

Farrugia, A. (2019). Australian Muslim population: How many Muslims in Australia? *New Idea*, 24 April.https://www.newidea.com.au/australian-muslim-population-how-many-muslims-in-australia.

Fekete, L. (2009). *A Suitable Enemy: Racism, migration and Islamophobia in Europe*. Pluto Press.

Feltz, V. (2006). Veils? Straw knows Jack. *The Daily Star*, 12 October..

Fenton, S., Carter, J., & Modood, T. (2000). Ethnicity and academia: Closure models, racism models and market models. *Sociological Research Online*, 5(2): 1–19.

Ferguson, J. (2009). The uses of neoliberalism. *Antipode*, 41: 166–184.

Fernando, M. L. (2009). Exceptional citizens: Secular Muslim women and the politics of difference in France. *Social Anthropology*, 17: 379–392.

Fernando, M. L. (2014). *The Republic Unsettled: Muslim French and the contradictions of secularism*. Duke University Press.

Fish, C. (2001). Condemnation without absolutes. *New York Times*, Opinion, 15 October.

Forrest, J., & Dunn, K. (2007). Constructing racism in Sydney, Australia's largest EthniCity. *Urban Studies*, 44(4): 699–721.

Forrest, J., & Dunn, K. (2010). Attitudes to multicultural values in diverse spaces in Australia's immigrant cities, Sydney and Melbourne. *Space and Polity*, 14(1): 81–102.

Forrest, J., & Dunn, K. (2011). Attitudes to diversity: New perspectives on the ethnic geography of Brisbane, Australia. *Australian Geographer*, 42(4): 435–354.

Fossé-Poliak, C., & Mauger, G. (1985). Choix politiques et choix de recherches: Essai d'auto-socio-analyse (1973–1984), *Cahiers du réseau Jeunesses et Sociétés*, 3–4–5: 27–121.

Foucault, M. (1980). Truth and power. In C. Gordon (ed.), *Michel Foucault Power/Knowledge: Selected interviews and other writings 1972–1977*, pp. 78–109. Pearson.

Foucault, M. (1991). Governmentality. In C. Gordon, G. Burchell, & P. Mille (eds), *The Foucault Effect: Studies in governmentality*. University of Chicago Press.

Frankenberg, R. (1993). Growing up white: Feminism, racism and the social geography of childhood. *Feminist Review*, 45(1): 51–84.

Freire, P. (1977). *Pédagogie des opprimés*, Maspero.

Friedrichs, J., Galster, G., & Musterd, S. (2003). Neighbourhood effects on social opportunities: The European and American research and policy context. *Housing Studies*, 18 (6): 797–806.

Fritzsche, L., & Nelson L. (2020). Refugee resettlement, place, and the politics of Islamophobia. *Social & Cultural Geography*, 21(4): 508–526.

Fyfe, N. R. (1998). Introduction: Reading the street. In N. R. Fyfe (ed.), *Images of the Street: Planning, identity and control in public space*, pp. 1–12. Routledge.

Gale, R. (2007). The place of Islam in the geography of religion: Trends and intersections. *Geography Compass*, 1(5): 1015–1036.

Gale, R. (2009). The multicultural city and the politics of religious architecture: Urban planning, mosques and meaning-making in Birmingham. In P. Hopkins & R. Gale (eds), *Muslims in Britain*, pp. 113–131. Edinburgh University Press.

Gale, R. (2013). Religious residential segregation and internal migration: The British Muslim case. *Environment and Planning A: Economy and Space*, 45: 872–891.

Gale, R., & Hopkins, P. (2009). Introduction: Muslims in Britain – Race, place and the spatiality of identities. In P. Hopkins & R. Gale (eds), *Muslims in Britain*, pp. 1–22. Edinburgh University Press.

Garbin, D. (2011). State-space: La Courneuve and Paris. In G. Millington (ed.), *Race, Culture and the Right to the City*, pp. 159–180. Palgrave Macmillan.

Garland, D. (2008). On the concept of moral panic. *Crime, Media, Culture*, 4: 9–30.

Garner, S. (2007). *Whiteness: An introduction*. Routledge.

Garner, S., & Selod, S. (2015). The racialization of Muslims: Empirical studies of Islamophobia. *Critical Sociology*, 41(1): 9–19.

Gartet F., & Id YassineR. (2013). Sociographie des lieux de culte musulman de Perpignan. In F. Dejean & L. Endelstein (eds), *Approches spatiales des faits religieux, Carnets de géographes*, p. 19, 6.

Gest, J. (2010). *Apart. Alienated and engaged Muslims in the West*. C. Hurst.

Ghani, H., & Nagdee, I. (2019). Islamophobia in UK universities. In I. Zempi, & I. Awan (eds), *The Routledge International Handbook of Islamophobia*, pp. 188–197. Routledge.

Ghumman, S., Ryan, A.M., Barclay, L.A., & Markel, K.S. (2013). Religious discrimination in the workplace: A review and examination of current and future trends. *Journal of Business and Psychology*, 28(4): 439–454.

Gieseking, J., & Mangold, W. (2014). *The People, Place and Space Reader*. Routledge.

Gilman, S. L. (2000). Are Jews white? In L. Back & J. Solomos (eds), *Theories of Race and Racism: A reader*, pp. 229–237. Routledge.

Ginther, D., & Kahn, S. (2009). Does science promote women? Evidence from academia 1973–2001. In R. Freeman & D. Goroff (eds), *Science and Engineering Careers in the United States: An analysis of markets and employment*, 163–194. University of Chicago Press.

Gintrac, C. (2012). Géographie critique, géographie radicale: Comment nommer la géographie engage. *Carnet de géographes*, 4: 1–13.

Gintrac, C. (2015). Quels positionnements pour quelles géographies critiques?. In A. Clerval et al (eds), *Espace et rapports de domination*, pp. 57–67. Presses Universitaires de Rennes.

Gintrac, C. (2017). La fabrique de la géographie urbaine critique et radicalex. *EchoGéo*, 39.

Githens-Mazer, J. and Lambert, R. (2010). *Islamophobia and Anti-Muslim Hate Crime: A London case study*. European Muslim Research Centre, University of Exeter.

Glynn, S. (2009). Liberalizing Islam: Creating Brits of the Islamic persuasion. In R. Phillips (ed.), *Muslim Spaces of Hope*, pp. 179–197. Zed Books.

Goetz, A. R., Vowles, T.M., & Tierney, S. (2009). Bridging the qualitative–quantitative divide in transport geography. *Professional Geographer*, 61(3): 323–335.

Gökariksel, B. (2012). The intimate politics of secularism and the headscarf: The mall, the neighbourhood, and the public square in Istanbul. *Gender, Place & Culture*, 19(1): 1–20.

Gökariksel, B., & Mitchell, K. (2005). Veiling, secularism, and the neoliberal subject: National narratives and supranational desires in Turkey and France. *Global Networks*, 5 (2): 147–165.

Gökarıksel, B., & Secor A. (2012). 'Even I was tempted': The moral ambivalence and ethical practice of veiling-fashion in Turkey. *Annals of the Association of American Geographers*, 102: 847–862.

Gökarıksel, B., & Secor A. (2015). Post-secular geographies and the problem of pluralism: Religion and everyday life in Istanbul, Turkey. *Political Geography*, 46: 21–30.

Göle, N. (2003). *Musulmanes et Modernes. Voile et civilisation en Turquie*. La Découverte.

Gould, P. (1999). *Becoming a Geographer*. Syracuse University Press.

Grafmeyer, Y., & Authier, J. Y. (2011). *Sociologie urbaine*. Colin.

Grasland, C. (2012). Le chercheur et le militant: réflexions 'à chaud' sur l'installation d'une centrale d'enrobés à Bonneuil-sur-Marne, *Métropolitiques*. 22 June.http://www.metrop olitiques.eu/Le-chercheur-et-le-militant.html.

Gregory, D., Johnston, R., Pratt, G., Watts, M., & Whatmore, S. (eds). (2011). *The Dictionary of Human Geography*. John Wiley & Sons.

Gregson, N., & Lowe, M. (1995). Home-making: On the spatiality of social reproduction in contemporary middle-class Britain. *Transaction of the Institute of British Geographers*, 20: 224–235.

Gresh, A. (2004). A propos de l'islamophobie. *Oumma*, 19 February.https://oumma. com/a-propos-de-lislamophobie/.

Grosfoguel, R. (2011). Decolonizing post-colonial studies and paradigms of political-economy: Transmodernity, decolonial thinking and global coloniality. *Transmodernity: Journal of Peripheral Cultural Production of the Luso-Hispanic World*, 1(1): 1–37.

Grosfoguel, R., & Mielants, E. (2012). The multiple faces of Islamophobia. *Islamophobia Studies Journal*, 1(1): 9–33.

Hackett, C. (2017). Five facts about the Muslim population in Europe. *Pew Research Center, Fact Tank*, 29 November.https://www.pewresearch.org/fact-tank/2017/11/ 29/5-facts-about-the-muslim-population-in-europe/.

Hafez, F. (2016). Comparing anti-Semitism and Islamophobia: The state of the field. *Islamophobia Studies Journal*, 3(2).

Hafez, F. (2018). Schools of thought in Islamophobia Studies: Prejudice, racism and decoloniality. *Islamophobia Studies Journal*, 4(2): 210–225.

Hage, G. (1998). *White Nation*. Pluto.

Hägerstrand, T. (1970). What about people in regional science? *Papers of the Regional Science Association*, 24: 7–21.

Hajjat, A. (2020). Islamophobia and French academia. *Current Sociology*. doi:10.1177/ 0011392120948920.

Halliday, F. (2003). *Islam and the Myth of Confrontation: Religion and politics in the Middle East*. I. B. Tauris.

Hamid, S. (2011). British Muslim young people: Facts, features and religious trends. *Religion, State and Society*, 39(2–3):247–261.

Hammersley, M. (2017). Deconstructing the qualitative–quantitative divide. In J. Brannen (ed.), *Mixing Methods: Qualitative and quantitative research*, pp. 3–37. Routledge.

Hammou, K., & Harchi, K. (2020). Nos plumes, nos voix. In O. Slaouti, & O. Le Cour Grandmaison (eds), *Racismes de France*, pp. 292–307. La Découverte, Cahiers libres.

Hancock, C. (2009). La justice au risque de la différence: Faire une 'juste place' à l'Autre. *Annales de géographie*, 61–75.

Hancock, C. (2013). Invisible Others: Muslims in European cities in the time of the burqa ban. *Treballs de la Societat Catalana de Geografia*, 75: 135–148.

Hancock, C. (2015). The Republic is lived with an uncovered face (and a skirt): (Un) dressing French citizens. *Gender, Place & Culture*, 22(7): 1023–1040.

Hancock, C. (2016). Traduttore traditore, The translator as traitor. *ACME: An International E-Journal for Critical Geographies*, 15(1): 15–35.

Hancock, C. (2017). Feminism from the margin: Challenging the Paris/banlieues divide. *Antipode*, 49(3): 636–656.

Hancock, C. (2020). Accommodating Islamophobia: How municipalities make place for Muslims in Paris. *Social & Cultural Geography*, 21(4): 527–545.

Hanson, S. (2010). Gender and mobility: New approaches for informing sustainability. *Gender, Place & Culture*, 17(1): 5–23.

Haraway, D. (1988). Situated knowledges: The science question in feminism and the privilege of partial perspective. *Feminist Studies*, 14(3): 575–599.

Harding, S. (1986). *The Science Question in Feminism*. Ithaca, NY: Cornell University Press.

Harding, S. (1998). *Is Science Multicultural? Postcolonialisms, feminisms and epistemologies*. Indiana University Press.

Harding, S. (2004a). Introduction: Standpoint theory as a site of political, philosophical, and scientific debate. In S. Harding (ed.), *The Feminist Standpoint Theory Reader: Intellectual and political controversies*, pp. 1–15. Psychology Press.

Harding S. (2004b). Rethinking standpoint epistemology: What is 'strong objectivity'?. In S. Harding (ed.), *The Feminist Standpoint Theory Reader: Intellectual and political controversies*, pp. 127–140. Routledge.

Hargreaves, A. G. (1996). A deviant construction: The French media and the 'banlieues'. *Journal of Ethnic and Migration Studies*, 22(4): 607–618.

Harris, C. D., & Ullman, E. L. (1945). The nature of cities. In P. K. Hatt & A. J. Reiss (eds), *Cities and Society: The revised reader in urban sociology*, pp. 237–247. Free Press of Glencoe.

Harrison, P. (2000). Making sense: Embodiment and the sensibilities of the everyday. *Environment and Planning D: Society and Space*, 18: 497–517.

Harvey, D. (1969). *Explanation in geography*. Edward Arnold.

Harvey, D. (1972). Revolutionary and counter-revolutionary theory in geography and the problem of ghetto formation. *Antipode*, 4(2): 1–13.

Harvey, D. (1973). *Social Justice and the City*. University of Georgia Press.

Harvey, D. (1976). The Marxian theory of the state. *Antipode*, 8(2): 80–89.

Harvey, D. (2000). *Spaces of Hope*. Blackwell.

Harvey, D. (2001). *Spaces of Capital: Towards a critical geography*. Routledge.

Harvey, D. (2007). *A Brief History of Neoliberalism*. Oxford University Press.

Harvey, D. (2010). *Géographie et capital: Vers un matérialisme historico-géographique*. Syllepse.

Harvey, D. (2011). *Le capitalisme contre le droit à la ville: Néolibéralisme, urbanisation, résistances.* Éditions Amsterdam.

Harvey, D. (2012). *Rebel Cities: From the right to the city to the urban revolution.* Verso.

Hassan, O. (2017). Trump, Islamophobia and US–Middle East relations. *Critical Studies on Security,* 5(2): 187–191.

Hedetoft, U., & Hjort M. (2002). Introduction. In U. Hedetoft & M. Hjort (eds), *The Postnational Self: Belonging and identity.* Minnesota: University of Minnesota Press.

Herzog, D. (2011). *Sexuality in Europe: A twentieth-century history.* Cambridge University Press.

Hewer, C. (2001). Schools for Muslims. *Oxford Review of Education,* 27(4): 515–527.

Hodge, D. (1995). Should women count? The role of quantitative methodology in feminist geographic research. *Professional Geographer,* 47(4): 426–426.

Holloway, L., & Hubbard P. (2001). *People and Place: The extraordinary geographies of everyday life.* Pearson Education Limited.

hooks, b. (1990). *Yearning: Race, gender, and cultural politics.* South End Press.

hooks, b. (2000). *Feminist Theory: From margin to center.* Pluto Press.

hooks, b. (2004). Choosing the Margin as a Space of Radical Openness. In S. Harding (ed.) *The feminist standpoint theory reader: Intellectual and political controversies,* pp. 153–159. Psychology Press.

hooks, b. (2008). Representing Whiteness in the Black imagination. In T. S. Oakes & P. L. Price (eds), *The Cultural Geography Reader,* pp. 373–379. Routledge.

Hopkins, P. (2004). Young Muslim men in Scotland: Inclusions and exclusions. *Children's Geographies,* 2(2): 257–272.

Hopkins, P. (2007). Global events, national politics, local lives: Young Muslim men in Scotland. *Environment and Planning A: Economy and Space,* 39(5): 1119–1133.

Hopkins, P. (2008). *The Issue of Masculine Identities for British Muslims after 9/11: A social analysis.* Edwin Mellen Press.

Hopkins, P. (2009). Geographical contributions to understanding contemporary Islam: Current trends and future directions. *Contemporary Islam,* 3(3): 213–227.

Hopkins, P. (2016). Gendering Islamophobia, racism and White supremacy: Gendered violence against those who look Muslim. *Dialogues in Human Geography,* 6(2): 186–189.

Hopkins, P. (2019). Social geography II: Islamophobia, transphobia, sizism. *Progress in Human Geography,* 030913251983347.

Hopkins, P., & Clayton, J. (2020). *Islamophobia and anti-Muslim hatred in North East England.* Tell MAMA/Newcastle University.

Hopkins, P., & Gale, R. (eds). (2009). *Muslims in Britain.* Edinburgh University Press.

Hopkins, P., & Smith, S. J. (2008). Scaling segregation: Racialising fear. In R. Pain, & S.J. Smith (eds), *Fear: Critical geopolitics and everyday life,* pp. 103–116. Ashgate.

Hoyt, H. (1939). *The Structure and Growth of Residential Neighbourhoods in American Cities.* Federal Housing Administration.

Hussain, A., & Yilmaz, I. (2012). Combatting terrorism in Britain: Choice for policy makers. In P. Weller & I. Yilmaz (eds), *European Muslims, Civility and Public Life: Perspectives on and from the Gülen Movement,* pp. 189–198. Continuum.

Husson, A-C. (2014). Une question de point de vue. *Genre!,* 15 April.https://cafaitgenre.org/2014/04/15/une-question-de-point-de-vue/.

Hwang, S. S., & Murdock, S.H. (1998). Racial attraction or racial avoidance in American suburbs? *Social Forces,* 77(2): 541–565.

Hyndman, J. (2004). Mind the gap: Bridging feminist and political geography through geopolitics. *Political Geography,* 23: 307–322.

Ibrahim, A. (2016). *The Rohingyas: Inside Myanmar's hidden genocide.* C. Hurst.

Iftikhar, A. (2020). India's new anti-Muslim law shows the broad allure of right-wing Islamophobic policies, *NBC News,* 13 January.https://www.nbcnews.com/think/op inion/india-s-new-anti-muslim-law-shows-broad-allure-right-ncna1112446.

Ignatieff, M. (1994). *Blood and Belonging: Journeys into the new nationalism.* Farrar, Straus and Giroux.

Ignatiev, N. (1995). *How the Irish became White.* Routledge.

Inayat, Q. (2007). Islamophobia and the therapeutic dialogue: Some reflections. *Counselling Psychology Quarterly,* 20(3): 287–293.

Iner, D., & Nebhan, K. (2019). Islamophobia from within: A case study on Australian Muslim women. In E. Bayrakli & F. Hafez (eds), *Islamophobia in Muslim Majority Societies,* pp. 199–215. Routledge.

Isakjee, A. (2016). Dissonant belongings: The evolving spatial identities of young Muslim men in the UK. *Environment and Planning A: Economy and Space,* 48(7): 1337–1353.

Isakjee, A., & Carroll B. (2019). Blood, body and belonging: The geographies of halal food consumption in the UK. *Social & Cultural Geography,* 1–22.

Isakjee, A., Davies, T., Obradović-Wochnik, J., & Augustová, K. (2020). Liberal violence and the racial borders of the European Union. *Antipode,* 52(6): 1751–1773.

Isin, E. F. (2002). *Being Political: Genealogies of citizenship.* University of Minnesota Press.

Isin, E. F. (2007). City, state: Critique of scalar thought. *Citizenship Studies,* 11: 211–228.

Itaoui, R. (2016). The geography of Islamophobia in Sydney: Mapping the spatial imaginaries of young Muslims. *Australian Geographer,* 47(3): 261–279.

Itaoui, R. (2020). Mapping perceptions of Islamophobia in the San Francisco Bay area, California. *Social & Cultural Geography,* 21(4): 479–506.

Itaoui, R., & Elsheikh, E. (2018). *Islamophobia in the United States: A reading resource pack.* Haas Institute for a Fair and Inclusive Society, University of California.

ITV Report. (2019). Government rejects controversial definition of Islamophobia after warning from terror police, *ITV News,* 19 May.https://www.itv.com/news/2019-05–15/poli ce-counter-terror-chief-joins-row-over-islamophobia-definition/.

Jackson, P. (ed.). (1987). *Race and Racism: Essays in social geography.* Unwin Hyman.

Jaggar, A. (2004). Feminist politics and epistemology: The standpoint of women. In S. Harding (ed.), *The Feminist Standpoint Theory Reader: Intellectual and political controversies,* pp. 55–80. Psychology Press.

Jargowsky, P-A. (1996). Take the money and run: Economic segregation in U.S. metropolitan areas. *American Sociological Review,* 61: 984–998.

Johnston, R. J. (2004). *Geography and Geographers: Anglo-American human geography since 1945.* Arnold.

Jones, T., & McEvoy, D. (1979). More on race and space. *Area,* 11(3): 222–223.

Joshi, S., McCutcheon, P., & Sweet, E. (2015). Visceral geographies of whiteness and invisible microaggressions. *ACME: An International E-Journal for Critical Geographies,* 14 (1): 298–323.

Kahera, A. (2010). *Deconstructing the American Mosque: Space, gender, and aesthetics.* University of Texas Press.

Kahera, A., Abdulmalik, L., & Anz, C. (2009). *Design Criteria for Mosques and Islamic Centres.* Routledge.

Kapoor, N. (2013). The advancement of racial neoliberalism in Britain. *Ethnic and Racial Studies,* 36(6): 1028–1046.

Khan, F., & Mythen, G. (2019). Micro-level management of Islamophobia: Negotiation, deflection and resistance. In I. Zempi & I. Awan (eds), *The Routledge International Handbook of Islamophobia*, pp. 313–324. Routledge.

Khattab, N., & Modood, T. (2015). Both ethnic and religious: explaining employment penalties across 14 ethno-religious groups in the United Kingdom. *Journal for the Scientific Study of Religion*, 54 (3).

Kilicbay, B., & Binark, M. (2002). Consumer culture, Islam and the politics of lifestyle. *European Journal of Communication*, 17(4): 495–511.

Kinsman, P. (1995). Landscape, race and national identity: The photography of Ingrid Pollard. *Area*, 27.

Klug, B. (2012). Islamophobia: A concept comes of ages. *Ethnicities*, 12(5): 665–681.

Kobayashi, A. (1994). Coloring the field: Gender, 'race', and the politics of fieldwork. *Professional Geographer*, 45(1): 73–80.

Kobayashi, A. (2006). Why women of colour in geography? *Gender, Place & Culture*, 13 (1): 33–38.

Kobayashi, A., & Peake, L. (2000). Racism out of place: Thoughts on whiteness and an antiracist geography in the new millennium. *Annals of the Association of American Geographers*, 90(2): 392–403.

Koefoed, L., & Simonsen, K. (2010). *'Den fremmede', byen og nationen: Om livet som etnisk minoritet*. ['The Stranger', the city and the nation: About living as an ethnic minority]. Roskilde universitets Forlag.

Kong, L. (2009). Situating Muslim geographies. In P. Hopkins & R. Gale (eds), *Muslims in Britain*, pp. 171–192. Edinburgh University Press.

Kong, L. (2010). Global shifts, theoretical shifts: Changing geographies of religion. *Progress in Human Geography*, 34(6): 755–776.

Kose, A. (1996). *Conversion to Islam: A study of native British converts*. Kegan Paul International.

Kruse, K. M. (2005). *White Flight: Atlanta and the making of modern conservatism*. Princeton: Princeton University Press.

Kumar, D. (2012). *Islamophobia and the Politics of Empire*. Haymarket.

Kundnani, A. (2014). *The Muslims Are Coming: Islamophobia, extremism and the domestic war on terror*. Verso.

Kwan, M-P. (1999). Gender and individual access to urban opportunities: A study using space–time measures. *Professional Geographer*, 51(2): 210–227.

Kwan, M.-P. (2004). Beyond difference: From canonical geography to hybrid geographies. *Annals of the Association of American Geographers*, 94: 756–763.

Kwan, M-P. (2008). From oral histories to visual narratives: Re-presenting the post-September 11 experiences of the Muslim women in the USA. *Social & Cultural Geography*, 9(6): 653–669.

Lacoste, Y. (1976). *La géographie, ça sert, d'abord, à faire la guerre*. Maspero.

Laurence, J., & Vaïsse, J. (2007). *Intégrer l'islam: la France et ses musulmans, enjeux et réussites*. Editions Odile Jacob.

Lawson, V. (1995). The politics of difference: Examining the quantitative/qualitative dualism in post-structuralist feminist research. *Professional Geographer*, 47: 449–457.

Le Goix, R. (2006). Les gated communities aux États-Unis et en France: une innovation dans le développement périurbain? *Hérodote*, 3: 107–137.

Lean, N. C. (2019). The debate over the utility and precision of the term 'Islamophobia'. In I. Zempi & I. Awan (eds), *The Routledge International Handbook of Islamophobia*, pp. 11–17. Routledge.

Lean, N. C., & Esposito, J. L. (2012). *The Islamophobia Industry: How the right manu-factures fear of Muslims*. Pluto.

Lee, T. (1978). Race, space and scale. *Area*, 10(6): 365–367.

Lefebvre, H. (1996). The right to the city. In *Writings on Cities*, trans. E. Kofman & E. Lebas (pp. 147–159). Blackwell.

Lentin, A. (and co-signatories) (2020). Open letter: The threat of academic authoritarian-ism - International solidarity with antiracist academics in France. *Open Democracy*, 5 November.https://www.opendemocracy.net/en/can-europe-make-it/open-letter-the-threat-of-academic-authoritarianism-international-solidarity-with-antiracist-academics-in-france/.

Lentin, A., & Titley, G. (2012). The crisis of 'multiculturalism' in Europe: Mediated min-arets, intolerable subjects. *European Journal of Cultural Studies*, 15(2): 123–138.

Leprince, C. (2019). Gloire, jeunesse et gros réseau: le recrutement au CNRS est-il arbitraire? *France Culture*, 19 June. https://www.franceculture.fr/sociologie/gloire-jeunesse-et-gros-reseau-le-recrutement-du-cnrs-est-il-arbitraire.

Letki, N. (2008). Does diversity erode social cohesion? Social capital and race in British neighbourhoods. *Political Studies*, 56: 99–126.

Lewis, R. (2009). Veils and sales: Muslims and the spaces of post-colonial fashion retail. In R. Phillips (ed.), *Muslim Spaces of Hope*, pp. 69–84. Zed Books.

Lipka, M. (2017). Muslims and Islam: Key findings in the U.S. and around the world. *Pew Research Center, Fact Tank*, 9 August.https://www.pewresearch.org/fact-tank/2017/08/09/muslims-and-islam-key-findings-in-the-u-s-and-around-the-world/.

Listerborn, C. (2015). Geographies of the veil: Violent encounters in urban public spaces in Malmö, Sweden. *Social & Cultural Geography*, 16: 95–115.

Little D. (2016). *Us Versus Them: The United States, Radical Islam, and the Rise of the Green Threat*. UNC Press Books.

Little, J. K. (2014). Society–space. In P. Cloke, P. Crang, & M. Goodwin. *Introducing Human Geographies*, 3rd edn, pp. 23–36. Abingdon: Routledge. ISBN: ISBN 9781444135350.

Livingstone, D. (1993). *The Geographical Tradition*. Blackwell.

Locke, J. (1689). *A Letter Concerning Toleration*. Awnsham Churchill.

Loewen, C. (2019). New Zealand mosque attacks leave Quebec Muslims feeling 'inde-scribable pain', *CBC News*, 15 March.https://www.cbc.ca/news/canada/ montreal/quebec-reax-new-zealand-mosque-shootings-1.5057737.

Lorcerie, F. (2005). *La Politisation du voile: L'affaire en France, en Europe et dans le monde arabe*. L'Harmattan.

Lorcerie, F., & Geisser, V. (2011). Muslims in Marseille. At Home in Europe project. *Open Society Foundations Report*, 322.

Losurdo, D. (2011). *Liberalism: A Counter-History*. Verso.

Luxembourg, C. (2016). David Harvey et la géographie radicale, *Mediapart, La Revue du Projet*, 11 May (54). https://blogs.mediapart.fr/edition/la-revue-du-projet/article/110516/david-harvey-et-la-geographie-radicale-corinne-luxembourg.

Mac an Ghaill, M., & Haywood, C. (2015). British-born Pakistani and Bangladeshi young men: Exploring unstable concepts of Muslim, Islamophobia and racialization. *Critical Sociology*, 41(1): 97–114.

Macdonald, M. (2006). Muslim women and the veil: Problems of image and voice in media representations. *Feminist Media Studies*, 6(1): 7–23.

MacMaster, N. (2012), *Burning the Veil: The Algerian War and the 'emancipation' of Muslim women, 1954–62*. Manchester University Press/Palgrave Macmillan.

Maddox, M. (2005). *God under Howard: The rise of the religious right in Australian politics*. Allen & Unwin.

Maddrell, A. (2012). Treasuring classic texts, engagement and the gender gap in the geographical canon. *Dialogues in Human Geography*, 2(3): 324–327.

Maddrell A. (2015). To Read or Not to Read? The Politics of Overlooking Gender in the Geographical Canon. *Journal of Historical Geography*, 49: 31–38.

Mahtani, M. (2006). Challenging the ivory tower: Proposing anti-racist geographies in the academy. *Gender, Place & Culture*, 13(1): 21–25.

Mahtani, M. (2014). Toxic geographies: Absences in critical race thought and practice in social and cultural geography. *Social & Cultural Geography*, 15(4): 359–367.

Maizland, L. (2019). China's repression of Uighurs in Xinjiang. *Council on Foreign Relations*, 25 November. https://www.cfr.org/backgrounder/chinas-repression-uighurs-xinjiang.

Majeed, S. (2019). Islamophobia and the mental health of Rohingya refugees. *Islamophobia and Psychiatry*, pp. 277–291. Springer.

Maldonado-Torres, N. (2007). On the coloniality of being. *Cultural Studies*, 21(2): 240–270.

Mamdani, M. (2004). *Good Muslim, Bad Muslim: America, the Cold War, and the roots of terror*. Pantheon.

Manji, I. (2004). *The Trouble with Islam: A Muslim's call for reform in her faith*. Random House.

Manski, C. (2000). Economic analysis of social interactions. *Journal of Economic Perspectives*, 14(3): 115–136.

Mansson McGinty, A. (2012). Teaching against culture in Geography of Islam. *Professional Geographer*, 64 (3): 358–369.

Mansson McGinty, A. (2014). Emotional geographies of veiling: The meanings of the hijab for five Palestinian American Muslim women. *Gender, Place & Culture*, 21: 683–700.

Mansson McGinty, A. (2020). Embodied Islamophobia: Lived experiences of anti-Muslim discourses and assaults in Milwaukee, Wisconsin. *Social & Cultural Geography*, 21(3): 402–420.

Mansson McGinty, A., Sziarto, K., & Seymour-Jorn, C. (2012). Researching within and against Islamophobia: A collaboration project with Muslim communities. *Social & Cultural Geography*, 14(1): 1–22.

Marx, K. (1844). Introduction to: A Contribution to the Critique of Hegel's Philosophy of Right. *Deutsch-Französische Jahrbücher*, 7.

Marx, K. (1967). *Capital*, 3 vols. International Publishers.

Marzouki, N. (2017). *Islam: An American religion*. Columbia University Press.

Mason, S. M. (2010). *Preterm birth risk in New York City's ethnic and immigrant enclaves*. Dissertation, University of North Carolina, Chapel Hill.

Massey, D. (1991). A global sense of place (first published in *Marxism Today*, June, pp. 24–29). In D. Gregory & T. Barnes, *Reading Human Geography: The poetics and politics of inquiry*. Arnold.

Massey, D. (2004). Segregation and stratification: A biosocial perspective. *Du Bois Review: Social Science Research on Race*, 1(1), 7–25.

Massey, D., & Denton, N. (1993). *American Apartheid: Segregation and the making of the underclass*. Harvard University Press.

Massey, J., & Tatla, R.S. (2012). Moral panic and media representation: The Bradford riot. In G. Morgan & S. Poynting (eds), *Global Islamophobia: Muslims and moral panic in the West*, pp. 161–180. Ashgate.

Massoumi, N., Mills, T., & Miller, D. (eds). (2017). *What is Islamophobia? Racism, social movements and the state*. Pluto.

Mayer, N., Tiberj, V., Vitale, T., & Michelat, G. (2019). Évolution et structures des préjugés: Le regard des chercheurs – section 1. Questions de méthode. *La lutte contre le racisme, l'antisémitisme et la xénophobie*, pp. 73–85, 2018.

McAuliffe, C. (2007). Visible minorities: Constructing and deconstructing the 'Muslim Iranian' diaspora. In C. Aitchison, P. Hopkins, & M-P. Kwan (eds), *Geographies of Muslim Identities*, pp. 29–55. Ashgate.

McAvay, H., &, Safi, M. (2018). Is there really such thing as immigrant spatial assimilation in France? Desegregation trends and inequality along ethnoracial lines. *Social Science Research*, 73: 45–62.

McDowell, L. (2008). Thinking through work: Complex inequalities, constructions of difference and trans-national migrants. *Progress in Human Geography*, 32(4): 491–507.

McIntosh, P. (1988). White privilege: Unpacking the invisible knapsack. *Peace and Freedom Magazine*, 10–12.

McKee, J. B. (1993). *Sociology and the Race Problem: The failure of a perspective*. University of Illinois Press.

McKendrick, J. H. (1996). Why bother with multi-method research? In *Multi-method Research in Population Geography: A Primer to Debate*, pp. 1–8. Population Geography Research Group of the Royal Geographical Society with the Institute of British Geographers.

McKittrick, K. (2011). On plantations, prisons, and a black sense of place. *Social & Cultural Geography*, 12(8): 947–963.

McKittrick, K., & Peake, L. (2005). What difference does difference make to geography. In N. Castree, A. Rogers, & D.J. Sherman (eds), *Questioning Geography: Fundamental debates*, pp. 39–54. Blackwell.

Meer, N. (2007). Muslim schools in Britain: Challenging mobilisations or logical developments? *Asia Pacific Journal of Education*, 27(1): 55–71.

Meer, N. (2008). The politics of voluntary and involuntary identities: Are Muslims in Britain an ethnic, racial or religious minority? *Patterns of Prejudice*, 42(1): 61–81.

Meer, N., & Modood, T. (2019). Islamophobia as the racialisation of Muslims. In I. Zempi, & I. Awan (eds), *The Routledge International Handbook of Islamophobia*, pp. 18–31. Routledge.

Mehmet, O. (1997). Al-Ghazzali on social justice: Guidelines for a new world order from an early medieval scholar. *International Journal of Social Economics*, 24(11): 1203–1218.

Mehta, U. (1999). *Liberalism and Empire: A study in nineteenth-century British liberal thought*. University of Chicago Press.

Merelli, A. (2019). Our failure to recognize past genocide makes it a present threat. *Quartz*, 17 December.https://qz.com/1770557/events-in-myanmar-and-india-show-genocide-is-a-current-threat/.

Merleau-Ponty, M. (1962). *Phenomenology of Perception*. Routledge & Kegan Paul.

Merriam, S., Johnson-Bailey, J., Lee, M., Kee, Y., Ntseane, G., & Muhamad, M. (2001). Power and positionality: Negotiating insider/outsider status within and across cultures. *International Journal of Lifelong Education*, 20(5): 405–416.

Merrifield, A. (1995). Situated knowledge through exploration: Reflections on Bunge's geographical expeditions. *Antipode*, 27(1): 49–70.

Merry, M. S. (2005). Advocacy and involvement: The role of parents in western Islamic schools. *Religious Education*, 100(4): 374–385.

Mills, C. W. (2004). Racial exploitation and the wages of whiteness. In G. Yancy (ed.), *What White Looks Like: African-American philosophers on the whiteness question*, pp. 25–55. Routledge. doi:10.4324/9780203499719.

Mills, S. (2009). Citizenship and faith: Muslim scout groups. In R. Phillips (ed.), *Muslim Spaces of Hope*, pp. 85–103. Zed Books.

Milton, K., & Svasek, M. (2005). *Mixed Emotions: Anthropological studies of feeling*. Berg.

Mir, S. (2014). *Muslim American Women on Campus: Undergraduate social life and identity*. UNC Press Books.

Mir, S., & Sarroub, L. K. (2019). Islamophobia in US education. In I. Zempi & I. Awan (eds), *The Routledge International Handbook of Islamophobia*, pp. 298–309. Routledge.

Mirza, H. S. (2006). Transcendence over diversity: Black women in the academy. *Policy Futures in Education*, 4(2): 101–113.

Mirza, H. S. (2013). 'A second skin': Embodied intersectionality, transnationalism and narratives of identity and belonging among Muslim in Britain. *Women's Studies International Forum*, 36: 5–15.

Modood. T. (1997). Introduction: The politics of multiculturalism in the New Europe. In T. Modood & P. Webner (eds), *The Politics of Multiculturalism in the New Europe*, pp. 1–26. Zed Books.

Modood, T. (2007). *Multiculturalism: A civic idea*. Polity Press.

Modood, T. (2008). Is multiculturalism dead? *Public Policy Research*, 15: 84–88.

Modood, T. (2009). Muslims and the politics of difference. In P. Hopkins & R. Gale (eds), *Muslims in Britain*, pp. 193–209. Edinburgh University Press.

Modood T., Berthoud R., Lakey J., Nazroo J., Smith P., Virdee S., & Beishons S. (eds.). (1997). *Ethnic Minorities in Britain: Diversity and disadvantage*. Policy Studies Institute.

Mohammad, R. (2005). Negotiating spaces of the home, the education system and the labour market. In G. Faleh, & C. Nagel, *Geographies of Muslim Women: Gender, religion and space*, pp. 178–202. Guilford Press.

Mohanty, C.T. (2003). *Feminism without Borders: Decolonizing theory, practicing solidarity*. Duke University Press.

Mondoloni, M. (2019). Soupçons de discriminations au CNRS: 200 universitaires dénoncent un 'acharnement' contre un candidat recalé trois fois, *Franceinfo*, 20 June.https://www.francetvinfo.fr/sciences/soupcons-de-discriminations-au-cnrs-200universitaires-de noncent-un-acharnement-contre-un-candidat-recale-trois-fois_3497741.html.

Mondon, A. (2015). The French secular hypocrisy: The extreme right, the Republic and the battle for hegemony. *Patterns of Prejudice*, 49(4): 392–413.

Mondon, A., & Winter, A. (2017). Articulations of Islamophobia: From the extreme to the mainstream? *Ethnic and Racial Studies*, 40(13): 2151–2179.

Mondon, A., & Winter, A. (2020). *Reactionary Democracy: How racism and the populist far right became mainstream*. Verso.

Monk, J. (2006). Changing expectations and institutions: American women geographers in the 1970s. *Geographical Review*, 96(2): 259–277.

Monk J., & Hanson, S. (1982). On not Excluding Half of the Human in Human Geography. *Professional Geographer*, 34(1): 11–23.

Moors, A. (2019). The trouble with transparency: Reconnecting ethics, integrity, epistemology, and power. *Ethnography*, 20(2), 149–169.

Moosavi, L. (2011). Muslim converts and Islamophobia in Britain. In T. Keskin (ed.), *The Sociology of Islam: Secularism, economy and politics*, pp. 247–268. Ithaca.

Moosavi, L. (2015). The racialization of Muslim converts in Britain and their experiences of Islamophobia. *Critical Sociology*, 41(1): 41–56.

Morelle, M., & Ripoll, F. (2009). Les chercheur-es face aux injustices: l'enquête de terrain comme épreuve éthique. *Annales de Géographie*, 665–666(1): 157–168.

Morgan, G., & Poynting, S. (2012). *Global Islamophobia: Muslims and moral panic in the West*. Ashgate.

Mott, C., & Cockayne, D. (2017). Citation matters: Mobilizing the politics of citation toward a practice of 'conscientious engagement'. *Gender, Place & Culture*, 24(7): 954–973.

Mott, C., & Cockayne, D. (2018). Conscientious disengagement and whiteness as a condition of dialogue. *Dialogues in Human Geography*, 8(2): 143–147.

Mudde, C. (2007). *Populist Radical Right Parties in Europe*. Cambridge University Press.

Mukherjee, S. R. (2015). Marianne voilée. *Histoire, monde et cultures religieuses*, 2: 83–107.

Musterd, S. (2003). Segregation and integration: A contested relationship. *Journal of Ethnic and Migration Studies*, 29(4): 623–641.

Musterd, S., & De Winter, M. (1998). Conditions for spatial segregation: Some European perspectives. *International Journal of Urban and Regional Research*, 22(4): 665–673.

Musterd S., & Ostendorf, W. (2005). Social exclusion, segregation and neighbourhood effects. In Y. Kazepov (ed.), *Cities of Europe: Changing contexts, local arrangements and the challenge to urban cohesion*, pp. 170–189. Blackwell.

Mythen, G., Walklate, S., & Khan, F. (2009). I'm a Muslim but I'm not a terrorist': Victimization, risky identities and the performance of safety. *British Journal of Criminology*, 49(6): 736–754.

Naber, N. (2008). Introduction. Arab Americans and US Racial Formations. In A. Jamal & N. Naber (eds), *Race and Arab Americans before and after 9/11: From invisible citizens to visible subjects*, pp. 1–45. Syracuse University Press.

Nagel, C. (2001). Contemporary scholarship and the demystification – and re mystification of 'Muslim women'. *Arab World Geographer*, 4(1): 63–72.

Nagel, C. (2016). Southern hospitality? Islamophobia and the politicization of refugees in South Carolina during the 2016 election season. *Southeastern Geographer*, 56(3): 283–290.

Nagel, C., & Staeheli, L. (2009). British Arab perspectives on religion, politics and the public. In P. Hopkins, & R. Gale (eds), *Muslims in Britain*, pp. 95–112. Edinburgh University Press.

Najib, K. (2013). *Dynamiques socio-spatiales et modes d'habiter des espaces urbains*. PhD thesis, University of Franche-Comté, Besançon, France.

Najib, K. (2018). Interdependence evaluation between the home neighborhood and the city: How socio-spatial categorization impacts residential segregation? *Social Sciences*, 7 (10): 178. doi:10.3390/socsci7100178.

Najib, K. (2019). Géographie et intersectionnalité des actes antimusulmans en région parisienne, Hommes & Migrations, January-March. *Religion and Discrimination*, 1324: 19–26.

Najib, K. (2020a). Spaces of Islamophobia and spaces of inequality in Greater Paris. *Environment and Planning C: Politics and Space*. doi:10.1177/2399654420941520.

Najib, K. (2020b). Socio-spatial inequalities and dynamics of rich and poor enclaves in three French cities: A policy of social mixing under test. *Population, Space and Place*, 26 (1): e2280. doi:10.1002/psp.2280.

Najib, K., & Finlay, R. (2020). Religion. In R. Pain & P. Hopkins (eds), *Social Geographies: An introduction*, pp. 143–151. Newcastle Social Geographies Collective. Rowman and Littlefield.

Najib, K., & Hopkins, P. (2019). Veiled Muslim women's strategies in responses to Islamophobia in Paris, *Political Geography*, 73: 103–111. doi:10.1016/j.polgeo.2019.05.005.

Najib, K., & Hopkins, P. (2020). Where does Islamophobia take place and who is involved? Reflections from Paris and London. *Social & Cultural Geography*, 21(4): 458–478. doi:10.1080/14649365.2018.1563800.

Najib, K., & Teeple Hopkins, C. (2020). Geographies of Islamophobia. *Social & Cultural Geography*, 21(4): 449–457. doi:10.1080/14649365.2019.1705993.

Naqshbandi, M. (2006). *Islam and Muslims in Britain: A guide for non-Muslims. Diversity in Action*. City of London Police, New Scotland Yard.

Naudier, D., & Simonet, M. (2011). Introduction. In D. Naudier et al. (eds), *Des sociologues sans qualité?*, pp. 5–21. La Découverte.

Nayak A. (2011). Geography, race and emotions: social and cultural intersections. *Social & Cultural Geography*, 12(6): 548–562.

Naylor, S., & Ryan, J.R. (2002). The mosque in the suburbs: Negotiating religion and ethnicity in South London. *Social & Cultural Geography*, 3: 39–59.

Neal, S., Agyeman, J. (2006). *The New Countryside? Ethnicity, nation and exclusion in contemporary rural Britain*. Policy Press.

Nelson, J., & Dunn, K. (2017). Neoliberal anti-racism: Responding to 'everywhere but different' racism. *Progress in Human Geography*, 41(1): 26–43.

Newman, D. (1985). Integration and ethnic spatial concentration: The changing distribution of the Anglo-Jewish community. *Transactions of the Institute of British Geographers*, 10(3): 360–376. doi:10.2307/622184.

Noble, G. (2005). The discomfort of strangers: Racism, incivility and ontological security in a relaxed and comfortable nation. *Journal of Intercultural Studies*, 26(1): 107–120.

Noble, G., & Poynting, S. (2008). Neither relaxed nor comfortable: The affective regulation of migrant belonging in Australia. In R. Pain & S.J. Smith (eds), *Fear: Critical geopolitics and everyday life*, pp. 129–138. Ashgate.

Noble, G., & Poynting, S. (2010). White lines: The intercultural politics of everyday movement in social spaces. *Journal of Intercultural Studies*, 31: 489–505.

Nora, P. (1987). *Essais d'ego-histoire*. Gallimard.

Noxolo, P. (2017) Introduction: Decolonising geographical knowledge in a colonised and re-colonising postcolonial world. *Area*, 49(3): 317–319.

Noxolo, P., Raghuram, P., & Madge, C. (2008). 'Geography is pregnant' and 'Geography's milk is flowing': Metaphors for a postcolonial discipline? *Environment and Planning D: Society and Space*, 26: 146–168.

Nyborg, K. (2000). Homo economicus and Homo politicus: Interpretation and aggregation of environmental values. *Journal of Economic Behavior & Organization*, 42(3): 305–322.

Oliver C., & Morris A. (2020). (Dis-)belonging bodies: Negotiating outsider-ness at academic conferences. *Gender, Place & Culture*, 27(6): 765–787.

Openshaw, S. (1998). Towards a more computationally minded scientific human geography. *Environment and Planning A: Economy and Space*, 30: 317–332.

Openshaw, S., & Taylor, P-J. (1979). A million or so – correlation coefficients: Three experiments on the modifiable areal unit problem. In N. Wrigley (ed.), *Statistical Methods in Spatial Sciences*, pp. 127–144. Pion.

OSF. (2011). *Unveiling the Truth: Why 32 women wear the full-face veil in France* (researched and written by Naïma Bouteldja). Open Society Foundations Report.

Osman, M. N. M. (2017). Understanding Islamophobia in Asia: The cases of Myanmar and Malaysia. *Islamophobia Studies Journal*, 4(1): 17–36.

Osman, M. N. M. (2019). Understanding Islamophobia in southeast Asia. In I. Zempi & I. Awan (eds), *The Routledge International Handbook of Islamophobia*, pp. 286–297. London and Routledge.

Oxfam. (2018). Reward Work, Not Wealth. *Oxfam briefing paper*. Oxfam. doi:10.21201/2017.1350.

Oza, R. (2007). Contrapuntal geographies of threat and insecurity: The United States, India and Israel. *Environment and Planning D: Society and Space*, 25(1): 9–31.

Pager, D., & Shepherd, H. (2008). The sociology of discrimination: Racial discrimination in employment, housing, credit, and consumer markets. *Annual Review of Sociology*, 34: 181–209.

Pain, R. (1991). Space, sexual violence and social control: integrating geographical and feminist analyses of women's fear of crime. *Progress in Human Geography*, 15: 415–431.

Pain, R. (2001). Gender, race, age and fear in the city. *Urban Studies*, 38(5–6): 899–913.

Pain, R. (2009). Globalized fear? Towards an emotional geopolitics. *Progress in Human Geography*, 33 (4): 466–486.

Pain, R., Panelli, R., Kindon, S., & Little, J. (2010). Moments in everyday/distant geopolitics: Young people's fears and hopes. *Geoforum*, 41: 972–982.

Pain, R., & Smith, S. (2008). *Fear: Critical geopolitical and everyday life*. Ashgate.

Pain, R., & Staeheli, L. (2014). Introduction: Intimacy-geopolitics and violence. *Area*, 46: 344–347.

Pande, R. (2020). Radical geographies. In H. Wilson & J. Darling (eds), *Research Ethics for Human Geography: A handbook for students*, pp. 107–117. Sage.

Panelli, R. (2008). Social geographies: Encounters with indigenous and more-than-White/Anglo geographies. *Progress in Human Geography*, 32(6): 801–811.

Pan Ké Shon, J-L. (2007). Portrait statistique des zones urbaines sensibles. *Informations sociales*, 5 (141): 24–32.

Parekh, B. (1992). The cultural particularity of liberal democracy. *Political Studies*, 40(1): 160–175.

Parekh, B. (2000). *The Future of Multi-ethnic Britain*. Profile Books.

Parker, I. (1994). Reflexive research and the grounding of analysis: Social psychology and the psy-complex. *Journal of Community & Applied Social Psychology*, 4: 239–252.

Paterson, J. L. (2014). *David Harvey's Geography (RLE Social & Cultural Geography)*. Routledge.

Paterson, J. L., Walters, M.A., & Brown, R. (2019). 'Your pain is my pain': Examining the community impacts of Islamophobic hate crimes. In I. Zempi & I. Awan (eds), *The Routledge International Handbook of Islamophobia*, pp. 84–96. Routledge.

Patton, T. (2004). Reflections of a Black woman professor: Racism and sexism in academia. *Howard Journal of Communications*, 15(3): 185–200.

Peach, C. (1996). Does Britain have ghettos? *Transactions of the Institute of British Geographers*, 21: 216–235.

Peach, C. (2005). Britain's Muslim population: An overview. In T. Abbas (ed.), *Muslim Britain: Community under pressure*, pp. 18–30. Zed Books.

Peach, C. (2006a). Muslims in the 2001 Census of England and Wales: Gender and economic disadvantage. *Ethnic and Racial Studies*, 29(4): 629–655.

Peach, C. (2006b). Islam, ethnicity and South Asian religions in the London 2001 census. *Transactions of the Institute of British Geographers*, 31(3): 353–370.

Peach, C., & Gale, R. (2003). Muslims, Hindus, and Sikhs in the new religious landscape of England. *Geographical Review*, 93(4), 469–490.

Peake, L. (2009). Feminist and quantitative? Measuring the extent of domestic violence in Georgetown, Guyana. *Treballs de la Societat Catalana de Geografia*, 66: 133–149.

Peake, L., & Sheppard, E. (2014). The emergence of radical/critical geography within North America. *ACME: An International E-Journal for Critical Geographies*, 13(2): 305–327.

Pearson, H. (1957). The economy has no surplus: A critique of a Theory of Development. In K. Polanyi, C. Arensberg, & H. Pearson (eds), *Trade and Market in Early Empires*. Free Press.

Pearson, E., & Winterbotham, E. (2017). Women, gender and Daesh radicalisation. *RUSI Journal*, 162(3): 60–72.

Peek, L.A. (2003). Reactions and response: Muslim students experiences on New York City campuses post 9/11. *Journal of Muslim Minority Affairs*, 23(3): 271–283.

Peet, R. (1977). The development of radical geography in the United States. *Progress in Human Geography*, 1(2): 240–263.

Peet, R. (1998). *Modern Geographical Thought*. Blackwell.

Perry, B. (2014). Gendered Islamophobia: Hate crime against Muslim women. *Social Identities*, 20(1): 74–89.

Pew Research Centre. (2007). *Muslim Americans: Middle class and mostly mainstream*. www.pewresearch.org/pubs/483/muslim-americans.

Philip, L. J. (1998). Combining quantitative and qualitative approaches to social research in human geography – An impossible mixture? *Environment and Planning A: Economy and Space*, 30(2): 261–276.

Phillips, D. (2006). Parallel lives? Challenging discourses of British Muslim self-segregation. *Environment and Planning D: Society and Space*, 24(1): 25–40.

Phillips, D. (2009). Creating home spaces: Young British Muslim women's identity and conceptualisations of home. In P. Hopkins & R. Gale (eds), *Muslims in Britain*, pp. 23–36. Edinburgh University Press.

Phillips, R. (2009). *Muslims Spaces of Hope: Geographies of possibility in Britain and the west*. Zed Books.

Phillips, R., & Iqbal, J. (2009). Muslims and the anti-war movements. In R. Phillips (ed.), *Muslim Spaces of Hope*, pp. 163–178. Zed Books.

Pickerill, J. (2016). Radical geography. In W. Liu & R. Marston (eds), *International Encyclopedia of Geography: People, the Earth, Environment and Technology*, pp. 1–14. John Wiley.

Piketty, T. (2013). *Capital in the 21st Century*. Harvard: Harvard University Press.

Pinçon, M., & Pinçon-Charlot, M. (2014). *La violence des riches*. La Découverte.

Pollard, J., Lim H., & Brown R. (2009). Muslim economic initiatives: Global finance and local projects. In R. Phillips (ed.), *Muslim Spaces of Hope*, pp. 139–162. Zed Books.

Poole, E. (2002). *Reporting Islam: Media representations of British Muslim*. I. B. Taurus.

Poon, J. (2005). Quantitative methods: Not positively positivist. *Progress in Human Geography*, 29(6): 766–772.

Poynting, S., & Mason, V. (2007). The resistible rise of Islamophobia: Anti-muslim racism in the UK and Australia before 11 September 2001. *Journal of Sociology*, 43(1): 61–86.

Poynting, S., & Noble, G. (2004). *Living with Racism: The experience and reporting by Arab and Muslim Australians of discrimination, abuse and violence since 11 September 2001*. Report to the Human Rights and Equal Opportunity Commission. Sydney: Centre for Cultural Research, University of Western Sydney.

Poynting, S., & Perry, B. (2007). Climates of hate: Media and state inspired victimisation of Muslims in Canada and Australia since 9/11. *Current Issues in Criminal Justice*, 19(2): 151–171.

Preteceille, E. (2009). La ségrégation ethno-raciale a-t-elle augmenté dans la métropole Parisienne. *Revue Française de Sociologie*, 50(3): 489–519.

Pulido, L. (2002). Reflections on a white discipline. *Professional Geographer*, 54(1): 42–49.

Ramahi, D. A. (2020). *Conversion to Islam and family relations in contemporary Britain.* Doctoral dissertation, University of Cambridge.

Ramamurthy, A. (2013). *Black Star: Britain's Asian youth movements.* Pluto.

Ray, E., Debah, S., & Ghenania, R. (2014). *Report on Islamophobia in Europe: Description of a scourge.* FEMYSO, IMAN, CCIF.

Razack, S. (2008). *Casting Out: The eviction of Muslims from Western law and politics.* University of Toronto Press.

Renton, D. (2003). Examining the success of the British National Party, 1999–2003. *Race & Class*, 45(2): 75–85.

Rex, J., & Moore, R. (1967). *Race Community and Conflict.* Oxford University Press.

Riaz, M. N., & Chaudry, N.M. (2003). *Halal Food Production.* CRC Press.

Robertson, R. (1995). Glocalization: Time-space and homogeneity–heterogeneity. In M. Featherstone, S. Lash, and R. Robertson (eds), *Global Modernities*, pp. 25–44. Sage.

Rootham, E. (2014). Embodying Islam and laïcité: Young French Muslim women at work. *Gender, Place & Culture*, 22(7): 971–986.

Rosanvallon, P. (2015). *Le bon gouvernement.* Seuil.

Rose, E. J. B. (1969). *Colour and Citizenship: A report on British race relations.* Institute of Race Relations, with Oxford University Press.

Rose, G. (1993). *Feminism and Geography.* Polity Press.

Roy, O. (2013). What matters with conversion? In N. Marzouki & O. Roy (eds), *Religious Conversions in the Mediterranean World*, pp. 175–187. Palgrave Macmillan.

Roy, O. (2020). French battle against Islamist 'separatism' is at odds with commitment to liberty, *Financial Times*, 7 November. https://www.ft.com/content/1f2f66c9-ce39-47dd-a1e9-a8687ff8ee2c.

Runnymede Trust. (1997). *Islamophobia: A challenge for us all.* Runnymede Trust.

Runnymede Trust. (2017). *Islamophobia: Still a challenge for us all.* Runnymede Trust.

Sack, R. (2007). *Homo geographicus.* Johns Hopkins University Press.

Saeed, A. (2007). Media, racism and Islamophobia: The representation of Islam and Muslims in the media. *Sociology Compass*, 1: 443–462.

Saeed, T. (2016). *Islamophobia and Securitization: Religion, ethnicity and the female voice.* Springer.

Saeed, T., & Johnson D. (2016). Intelligence, global terrorism and higher education: Neutralising threats or alienating allies? *British Journal of Educational Studies*, 64(1): 37–51.

Safi, M. (2009). La dimension spatiale de l'intégration: Evolution de la segregation des populations immigrées en France entre 1968 et 1999. *Revue Française de Sociologie*, 50 (3): 521–552.

Said, E. (1978). *Orientalism.* Pantheon Books.

Saikia, A. (2020). I coloured my sword red: Meet Delhi rioters who say they killed Muslims. *Scroll News*, 11 March.https://scroll.in/article/955044/meet-the-rioters-who-sa y-they-killed-muslims-in-delhi-violence.

Salih, R. (2000). Shifting boundaries of self and other: Moroccan migrant women in Italy. *European Journal of Women's Studies*, 7(3): 321–335.

Samari, G. (2016). Islamophobia and public health in the United States. *American Journal of Public Health*, 106(11): 1920–1925.

Sardar, Z. (2009). Spaces of hope: Interventions. In R. Phillips (ed.), *Muslim Spaces of Hope*, pp. 13–26. Zed Books.

Sayyid, S. (2014). *Recalling the Caliphate: Decolonization, and world order*. C. Hurst.

Sayyid S., & Vakil A. (2010). *Thinking through Islamophobia: Global perspectives*. C. Hurst.

Scharff, C. (2011). Disarticulating feminism: Individualization, neoliberalism and the othering of 'Muslim women'. *European Journal of Women's Studies*, 18(2): 119–134.

Schelling, T. C. (1969). Models of segregation. *American Economic Review*, 59(2): 488–493.

Schelling, T. C. (1971). Dynamic models of segregation. *Journal of Mathematical Sociology*, 1: 143–186.

Schelling, T. C. (1978). *Micromotives and Macrobehavior*. WW Norton & Company.

Schiffer, S., & Wagner, C. (2011). Anti-Semitism and Islamophobia: New enemies, old patterns. *Race & Class*, 52(3): 77–84.

Schwartz, S. (2010). Islamophobia: America's new fear industry. *Phi Kappa Phi Forum*, 90 (3): 19–21.

Scott, H. (2003). Stranger danger: Explaining women's fear of crime. *Western Criminology Review*, 4(3).

Scott, W. J. (2007). *The Politics of the Veil*. Princeton University Press.

Secor, A. (2002). The veil and urban space in Istanbul: Women's dress, mobility and Islamic knowledge. *Gender, Place & Culture*, 9: 5–22.

Selod, H. (2005). La mixité sociale: le point de vue des sciences économiques. *Informations sociales*, 5(125): 28–35.

Selod, S. (2015). Citizenship denied: The racialization of Muslim American men and women post-9/11. *Critical Sociology*, 41(1): 77–95.

Serrant-Green, L. (2002). Black on black: Methodological issues for black researchers working in minority ethnic communities. *Nurse Researcher*, 9(4): 30–44.

Sharp, J. (2007). Geography and gender: Finding feminist political geographies. *Progress in Human Geography*, 31(3): 381–388.

Shaw, W. S., Herman, R. D., & Dobbs, G. R. (2006). Encountering indigeneity: Reimagining and decolonizing geography. *Geografiska Annaler*, 88B: 267–276.

Sheppard, E. (2001). Quantitative geography: Representations, practices, and possibilities. *Environment and Planning D: Society and Space*, 19: 535–554.

Sherman, D., Rogers, A., & Castree, N. (2005). Introduction: questioning geography. In N. Castree, A. Rogers, & D.J. Sherman (eds), *Questioning Geography: Fundamental debates*. Blackwell.

Shterin, M., & Spalek, B. (2011). Muslim young people in Britain and Russia: Intersections of biography, faith and history. *Religion, State & Society*, 39(2–3):145–154.

Sian, K. (2017). Being black in a white world: Understanding racism in British universities. Papeles del CEIC. *International Journal on Collective Identity Research*, 2: 2.

Simon, P., & Tiberj, V. (2013). Sécularisation ou regain religieux: la religion des immigrés et de leurs descendants. INED, Documents de travail. *Séries Trajectoires et Origines*, 196: 47.

Siraj, A. (2011). Meanings of modesty and the hijab amongst Muslim women in Glasgow, Scotland. *Gender, Place & Culture*, 18(6): 716–731.

Sirin, S. R., & Fahy, S. (2009). What do Muslims want? A voice from Britain. *Analyses of Social Issues and Public Policy*, 6(1).

Sirin, S. R., & Fine, M. (2007). Hyphenated selves: Muslim American youth negotiating identities on the fault lines of global conflict. *Applied Development Science*, 11(3): 151–163.

Sirin, S. R., & Imamoglu, S. (2009). Muslim-American hyphenated identity: Negotiating a positive path. In R. Phillips (ed.), *Muslim Spaces of Hope*, pp. 236–251. Zed Books.

Smith, N. (1992). Geography, difference and the politics of scale. In J. Doherty, E. Graham, & M. Malek (eds), *Postmodernism and the Social Sciences*, pp. 57–79. Macmillan.

Smith, S. (1989). *The Politics of 'Race' and Residence*. Polity.

Soja, E. (2010). *Postmodern Geographies: The reassertion of space in critical social theory*. Verso.

Speck, B. W. (1997). Respect for religious differences: The case of Muslim students. *New Directions for Teaching and Learning*, 70: 39–46.

Spivak, G. C. (1988). Can the subaltern speak? In C. Nelson & L. Grossberg (eds), *Marxism and the Interpretation of Culture*, pp. 271–313. University of Illinois Press.

Springer, S. (2016). *The anarchist roots of geography: Toward spatial emancipation*. University of Minnesota Press.

Staeheli, L. (1996). Publicity, privacy, and women's political action. *Environment and Planning D: Society and Space*, 14: 601–619.

Staeheli, L., & Nagel, C. (2008). Rethinking security: Perspectives from Arab-American and British Arab activists. *Antipode*, 40: 780–801.

Stanley, L., & Wise, S. (1983). *Breaking Out: Feminist consciousness and feminist research*. Routledge and Kegan Paul.

Staszak, J-F. (dir.). (2001). *Géographies anglo-saxonnes. Tendances contemporaines*. Belin.

Sue, D. W. (1993). Confronting ourselves: The White and racial/ethnic-minority researcher. *Counseling Psychologist*, 21(2): 244–249.

Sullivan, S. (2006). *Revealing Whiteness: The unconscious habits of racial privilege*. Indiana University Press.

Swanton, D. (2010). Sorting bodies: Race, affect, and everyday multiculture in a mill town in northern England. *Environment and Planning A: Economy and Space*, 42(10): 2332–2350.

Swanton, D. (2016). Encountering Keighley: More than human geographies of difference in a former mill town. In J. Darling & H. F. Wilson (eds), *Encountering the City: Urban encounters from Accra to New York*, pp. 111–130. Routledge.

Taeuber, K. E., & Taeuber, A.F. (1965). *Negroes in Cities: Residential segregation and neighborhood change*. Aldine.

Tajfel, H., & Turner J.C. (1986). The social identity theory of intergroup behavior. In S. Worchel, & W. G. Austin (eds), *Psychology of Intergroup Relations*, pp. 7–24. Nelson-Hall.

Taras, R. (2013). Islamophobia never stands still: Race, religion, and culture. *Ethnic and Racial Studies*, 36(3): 417–433.

Teeple Hopkins, C. (2015). Social reproduction in France: Religious dress laws and laïcité. *Journal of Women's Studies International Forum*, 48: 154–164.

Tell MAMA. (2016). Geography of anti-Muslim Hate in 2015. Annual report 2015. *Monitoring anti-Muslim attacks*, pp. 38–51. Faith Matters. https://www.tellmamauk. org/wp-content/uploads/pdf/tell_mama_2015_annual_report.pdf.

Tissot, S., & Poupeau, F. (2005). La spatialisation des problèmes sociaux. *Actes de la recherche en sciences sociales*, 4(159): 4–9.

Todd, E. (2015). *Qui est Charlie? Sociologie d'une crise religieuse*. Seuil.

Tolia-Kelly, D. (2017). A day in the life of a geographer: 'Lone', black, female. *Area*, 49: 324–328.

Tovar, E. (2011). Comment mesurer la ségrégation urbaine? Une contribution économique. *Cybergéo: European Journal of Geography*, article 548.

Twigger-Ross, C., & Uzzel, D. (1996). Place and identity process. *Journal of Environmental Psychology*, 16: 205–220.

Tyler, K. (2012). *Whiteness, Class and the Legacies of Empire: On home ground.* Basingstoke: Palgrave Macmillan.

Tyrer, D. (2003). *Institutionalized Islamophobia in British universities.* PhD thesis, University of Salford.

Tyrer, D., & Ahmad, F. (2006). *Muslim Women and Higher Education: Identities, experiences and prospects: A summary report.* John Moores University and European Social Fund.

United Nations. (2015). *World urbanization prospects: 2014 revision.* UN Department of Economics and Social Affairs, Population Division.

Valentine, G. (1989). The geography of women's fear. *Area,* 21: 385–390.

Valentine, G. (2001). *Social Geographies: Space and society.* Prentice Hall.

Valentine, G. (2008). Living with difference: Reflections on geographies of encounter. *Progress in Human Geography,* 32(3): 323–337.

Valentine, G., Skelton, T., & Chambers, D. (1998). Cool places: An introduction to youth and youth cultures. In T. Skelton & G. Valentine (eds), *Cool Places: Geographies of youth cultures.* Routledge.

Valins, O. (2003). Defending identities or segregating communities? Faith-based schooling and the UK Jewish community. *Geoforum,* 34(2): 235–247.

Van Nieuwkerk, K. (2004). Veils and wooden clogs don't go together. *Ethnos,* 69(2): 229–246.

Vieillard-Baron, H. (2004). De la difficulté à cerner les territoires du religieux: Le cas de l'islam en France. *Annales de Géographie,* 113(640): 563–587.

Voas, D., & Williamson, P. (2000) The scale of dissimilarity: Concepts, measurements and an application to socioeconomic variation across England and Wales. *Transactions of the Institute of British Geographers,* 25: 465–481.

Wacquant, L. (1993). Urban outcasts: Stigma and division in the Black American ghetto and the French urban periphery. *International Journal of Urban and Regional Research,* 17: 366–383.

Wacquant, L. (2007). *Urban Outcasts: A comparative sociology of advanced marginality.* Polity Press. ISBN: ISBN: 978-0-745-63125-7.

Wacquant, L. (2015). Revisiting territories of relegation: Class, ethnicity and state in the making of advanced marginality. *Urban Studies,* 53(6): 1077–1088.

Wakeling, P. (2007). White faces, black faces: Is British sociology a white discipline? *Sociology,* 41(5): 945–960.

Warren, S. (2019). #YourAverageMuslim: Ruptural geopolitics of British Muslim women's media and fashion. *Political Geography,* 69: 118–127.

Warrington, M. (2001). The geographies of domestic violence. *Transactions of the Institute of British Geographers,* 16: 265–282.

Watson, S. (2005). Symbolic space of difference: Contesting the Eruv in Barnet, London and Tenafly, New Jersey. *Environment and Planning D: Society and Space,* 23(4): 597–613.

Watt, P. (1998). Going out of town: 'Race' and place in the south east of England. *Environment and Planning D: Society and Space,* 16: 687–703.

Weber, M. (1959). *Le savant et le politique.* Plon.

Weller, P. (2006). Addressing religious discrimination and Islamophobia: Muslims and liberal democracies. The case of the United Kingdom. *Journal of Islamic Studies,* 17(3): 295–325.

Wilkinson, S. (1988). The role of reflexivity in feminist psychology. *Women's Studies International Forum,* 11(5): 493–502.

Williams, J., & Massaro, V. (2013). Feminist geopolitics: Unpacking (in)security, animating social change. *Geopolitics*, 18(4): 751–758.

Wilson, B. (1966). *Religion in Secular Society: A sociological comment*. Watts.

Wilson, H. F. (2016). Encountering Havana: Texts, aesthetics and documentary encounters. In J. Darling & H. F. Wilson (eds), *Encountering the City: Urban encounters from Accra to New York*, pp. 203–219. Routledge.

Wilson, H. F., & Darling J. (2016). The possibilities of encounter. In J. Darling & H. F. Wilson (eds), *Encountering the City: Urban encounters from Accra to New York*, pp. 1–24. Routledge.

Wilson, R. (1987). *The Truly Disadvantaged, the Inner City, the Underclass and Public Policy*. Chicago University Press.

Wirth, L. (1998). *The Ghetto*. Chicago University Press.

Wong L. (1994). Di(s)-ecting and dis(s)-closing 'Whiteness': Two tales about psychology. In K. Bhavnani & A. Phoenix (eds), *Shifting Identities: Shifting racisms: A feminism and psychology reader*, pp. 133–153. Sage.

Worth, O. (2014). The far-right and neoliberalism: Willing partner or hegemonic opponent? In R. Sall et al. (eds), *The Longue Durée of the Far-Right*, pp. 165–184. Routledge.

Young, I. (1990). *Justice and the Politics of Difference*. Princeton: Princeton University Press.

Younus, S., & Mian, A.I. (2018). Children, adolescents, and Islamophobia. In H. S. Moffic, J. Peteet, A. Z. Hankir, & R. Awaad (eds), *Islamophobia and Psychiatry: Recognition, prevention, and treatment*, pp. 321–334. Springer.

Yucel, S. (2010). *The Struggle of Ibrahim: Biography of an Australian Muslim*. Tughra.

Zempi, I., & Awan, I. (2019). *The Routledge International Handbook of Islamophobia*. Routledge.

Zempi, I., & Chakraborti, N. (2014). *Islamophobia, Victimisation and the Veil*. Palgrave Macmillan.

Zempi, I., & Chakraborti, N. (2015). 'They make us feel like we're a virus': The multiple impacts of Islamophobic hostility towards veiled Muslim women. *International Journal for Crime, Justice and Social Democracy*, 4(3): 44–56.

Zimmerman, D. D. (2015). Young Arab Muslim women's agency challenging Western feminism. *Affilia*, 30(2): 145–157.

Index

Spatialized Islamophobia (Najib)
ISBN: 978-0-367-89478-8 (hbk)
Page numbers in **bold** and *italic* type refer to information in tables and figures respectively.
Those followed by 'n' denote information in notes.

Printed in the United States
by Baker & Taylor Publisher Services

Printed in the United States
by Baker & Taylor Publisher Services